Weed Biology and Control
in Agriculture and Horticulture

David C. Gwynne
and
Robert B. Murray

Batsford Academic and Educational   London

© David C. Gwynne and Robert B. Murray 1985

First published 1985

Typeset by Deltatype, Ellesmere Port
and printed in Great Britain by
Biddles Ltd
Guildford and Kings Lynn

Published by Batsford Academic and Educational
an Imprint of B. T. Batsford Ltd
4 Fitzhardinge Street, London W1H 0AH

British Library Cataloguing in Publication Data

Gwynne, David C.
   Weed biology and control in agriculture
   and horticulture.
   1. Weeds——Control
   I. Title
   632'.58   SB611

   ISBN 0–7134–3531–3

Contents

Foreword

During the last 40 years the science of weed control has grown from a position of relative insignificance to become one of the major technologies of modern agriculture. Despite the high cost of developing new herbicides the number available to farmers and growers continues to increase. New herbicides with novel uses are becoming available while old herbicides are being used in new ways.

It is not surprising therefore that weed control recommendations for specific crops have changed significantly in the last decade. Selecting the most appropriate herbicide for a particular weed problem, crop, stage of growth, soil and climatic conditions has become a complex task made more difficult by the increased number of possible treatments. Concepts too are changing, not only because of continuing developments in herbicide technology, but also because of the growing recognition that herbicides are not merely weed killers but crop production chemicals.

Against this background regular sifting and revision of information is required. This book brings together in one inexpensive volume up-to-date information on the identification, biology and control of the weeds of temperate agriculture and horticulture.

The book has been aimed at diploma and degree students, personnel in local authorities, advisers, farmers and growers. The authors have many years' experience of teaching weed science on third level courses and of providing specialist advice on weed control. Their experience has enabled them to deal in simple terms with many relatively complex aspects of weed science such as herbicide formulations, selectivity and application methods.

In the use of herbicides label recommendations must be followed accurately; but such instructions cannot tell the whole story. The greater the understanding of underlying principles, the better will be the results achieved. This book will contribute greatly towards that end.

D. W. Robinson
Kinsealy Research Centre
Agricultural Institute
Dublin 5.

Preface

This book arose from a feeling engendered by many years' contact with students at all levels, and with farmers, growers and advisers, that such a general text would be a useful addition to the existing weed literature.

It should be emphasised at the outset that it has *not* been our intention to provide exhaustive coverage of individual herbicides or treatments. Details of dosage rates, timings and costs would be out of place, we feel, in a text of this type and are in any case readily available from MAFF and other publications (Appendix 3) and from manufacturers. Change and improvement in both chemicals and application technology are part and parcel of the herbicide scene and have as far as possible been documented in the text. In certain areas of interest, e.g. the introduction of the so-called graminicides, and the trend towards low-volume spraying, development is particularly rapid at the present time; the relevant literature, including the agricultural and horticultural press, should again be consulted for details.

References to published work take two forms. The Further Reading section at the end of each chapter includes major review articles and other items of comparable interest, while all other references have been given alphabetically in the Selected Bibliography at the end of the book, which is intended to provide a reasonably comprehensive entry to the more recent and accessible literature. Certain particularly valuable older items have also been included. A number of these references also appear in Appendix 3, which summarises the available information sources. *The Weed Control Handbook* (Roberts, 1982) provides background information on various aspects covered by the present text. A full, general, reference is given in the Selected Bibliography, while individual items of interest are included in the Further Reading sections, as appropriate.

References to specific herbicides are generally by their BSI common chemical name (e.g. simazine; mecoprop). Mention of proprietary products by brand name (indicated thus *) is not to be construed as an endorsement of these products, nor is omission of others to be taken as implied criticism. Although every effort has been made to ensure accuracy, the authors cannot accept responsibility for any loss, damage or other accident arising out of the use of procedures discussed in this text.

With regard to weed names, in order to allow the text to flow more freely the convention has been adopted of giving scientific (Latin) names only where a species is discussed in some detail, as for example in Section 1.1. A full English/Latin glossary is included in Appendix 4. English names are taken from *English Names of Wild Flowers*, by J. G. Dony, F. H. Perring and C. M. Rob (Butterworths, London, 1974) and Latin names from the *Flora of the British Isles*, by A. R. Clapham, T. G. Tutin and E. F. Warburg (C.U.P., 2nd edn, 1962) with some updating as in the *Excursion Flora of the British Isles*, by the same authors (3rd edn, 1981) – but see glossary for exceptions.

The authors are grateful to the West of Scotland Agricultural College for encouragement to write this book and to colleagues at the College and elsewhere for helpful comments. In particular we thank Dr D. Hall (W.S.A.C.) and Mr D. McNab (Sports Turf Services Ltd) for their contributions to Sections 3.2 and 11.1 respectively. Thanks are due also to all those whose publications have been consulted and especially to Blackwell Scientific Publications, Oxford; Dr P. Dubach (Ciba-Geigy Ltd, Basle); Dr S. R. Moss (Weed Research Organisation) and others for kind permission to reproduce figures. We are indebted to Mrs E. Dixon, who typed successive drafts with noteworthy speed and accuracy and to Miss P. Wilson, who prepared many of the figures for publication.

Finally, we are particularly grateful to Dr David Robinson for a wealth of advice and information given to one of us (R.B.M.) and for kindly agreeing to write the foreword to this book.

D. C. Gwynne
R. B. Murray
March, 1984

1 Weed Origins and Distribution

1.1 Development of the weed flora

1.1.1 PREHISTORIC TO MODERN TIMES

In any consideration of present-day weed problems two main features must be kept in mind. First, the weed flora is not static, but shows marked changes over the years; and second, it is essentially man-made, the end-product of centuries of increasingly intensive human occupation of the land (Godwin, 1960).

As a suitable baseline we may take the situation at the end of the last glaciation of Britain, some 10,000 years ago. Among the plant species present at that early date it is interesting to note a number which are now prominent weeds, indicating perhaps a preference for the open ground and disturbed, mineral-rich soils typical both of post-glacial and of agricultural environments (Table 1.1).

The next few thousand years saw a gradual improvement in the climate of Britain, culminating in the 'climatic optimum' between 5,500 BC and 3,200 BC when mean temperatures were some 2°C above those of today. This was reflected in a continuing influx of warmth-demanding species and in the replacement of grassland and heath by woodland, which latterly covered the entire country to a height well above the present tree-line. As might be expected, few new weeds entered the country at that time, while those already present were forced into peripheral 'refuge' areas, such as cliffs, screes, riverbanks and flood-channels and coastal habitats; some species may indeed have been eliminated, to be reintroduced at a later date. Also about this time (5,500 BC), the rising sea produced by melting of the ice-sheets finally inundated the land-connection with the rest of the continent, bringing plant immigration into Britain to a halt. One result of this has been to give Britain a less diverse weed flora than that of adjacent parts of Europe, a matter of some relief to agriculturalists!

The subsequent vegetational history of Britain is one of gradual reduction of the virgin forest in the face of increasing human pressure, allied to deteriorating climatic and soil conditions. A major factor in this was the arrival from about 3,500 BC onwards of Neolithic people with a cattle-raising/cereal-growing economy. The advent of these first farmers is marked in the pollen record by a phenomenon known as the 'elm decline', a marked and generally permanent reduction in

Table 1.1: *Weeds and ruderals* recorded in Britain before 8000 BC*

Arable Weeds:	Grassland weeds and ruderals:
Orache	Common knapweed
Mouse-ear chickweed	Plaintains (greater, hoary & ribwort)
Fat-hen	Silverweed
Hemp-nettle	Buttercups (meadow and creeping)
Cleavers	Common and sheep's sorrel
Knotgrass	Dandelion
Redshank	Common nettle
Wild pansy	Marsh thistle
Docks	Spear thistle
Chickweed	Common toadflax
Scentless mayweed	
Perennial sow-thistle	
Cornflower	
Annual knawel	

*Ruderals – plants essentially suited to colonising waste ground, roadsides, riverbanks and similar situations, but not usually penetrating arable land itself.

elm-tree pollen, now thought to be due mainly to the selective utilisation of elm as cattle fodder (Rackham, 1976). This is almost always accompanied by an increase in the pollen of grasses and cereals and of weeds of occupation such as bracken, stinging nettle and, especially, plantain (Fig. 1.1).

With their cereal seed these immigrant farmers also brought to Britain 'fellow-travellers' in the shape of the first true arable weeds, although the concept itself was, of course, unknown to them (Table 1.2). Indeed, it appears that for hundreds of years the weeds were harvested along with the crop, forming part of the 'poor man's porridge' well into medieval times. There is much evidence of the utilisation of weed species by prehistoric and later peoples, confirmed in somewhat macabre fashion by the presence of seeds of pale persicaria, black bindweed, fat hen, and others, along with barley and linseed, in the stomachs of sacrificial victims preserved in Danish peat-bogs (Glob, 1969).

The continuing trend of forest clearance increased greatly with the introduction of iron tools by Celtic peoples, from about 500 BC onwards. Initially, the process was largely confined to lighter soils in the lowlands and some upland areas, penetration on the less tractable soils of south-east England and the Midlands being an achievement mainly of the Anglo-Saxons, with their wheeled, mouldboard plough (Evans, 1976).

There is plentiful archaeological evidence for the development of cereal husbandry in Iron Age and Roman times, however, and it is

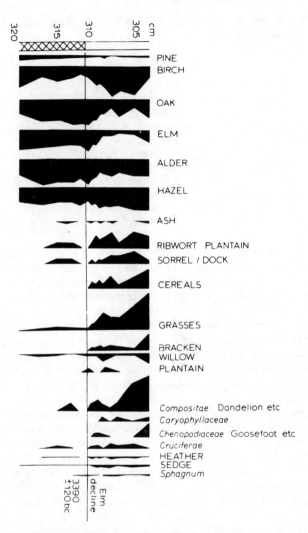

Fig. 1.1: *Pollen diagram, Barfield Tarn, Cumbria; the elm decline and associated increase in grass, cereal and weed pollen (after Evans, 1976).*

known that corn was regularly exported from Britain to the Continent during the latter part of this period. In the course of this trade there was a renewed influx of weed species, and various authors (e.g. Salisbury, 1961) have noted the significance of the extensive Roman road system as a focus for dispersal. Many weed species increased markedly at this time, including several whose association with

Table 1.2: *Weeds first recorded in Britain from the Neolithic and Bronze Ages (c. 3,500 BC–c.500 BC)*

Common poppy
Charlock
Field pennycress
Black-bindweed
Smooth tare
Ivy-leaved speedwell
Red dead-nettle
Rye brome (chess)
Barren brome
Onion-couch
Field madder
Black medick
Creeping thistle

important Iron Age crops such as cereals (e.g. darnel; wild-oats) and flax (e.g. spurrey; pennycress) is well established (Table 1.3).

It is now recognised that despite the major social and other upheavals of the centuries following the Roman withdrawal, life at 'grass-roots' level went on in many respects much as before. The Anglo-Saxons completed in large measure the destruction of the primeval forest in the English lowlands, and extended agriculture into the fens, midland valleys and other heavy land areas. In other parts of lowland Britain suitable documentary evidence is more scarce, but what there is indicates a similar situation by the seventeenth century at the latest.

The development of the weed flora can be seen, therefore, as an ongoing process, with the expanding arable acreage offering increasing scope for weed establishment. Few new introductions were recorded during the latter part of this period, which saw a gradual sorting-out of weed distributions on the basis of ecological and crop/husbandry tolerances and which eventually produced what came to be known as the 'traditional' weed-flora of lowland Britain.

1.1.2 THE AGRICULTURAL REVOLUTION – 1700 TO 1900

The history of these plants, many of which have been largely swept away by more recent events, is documented in detail by Salisbury (1961) and only a few typical examples are discussed here. Darnel (*Lolium temultentum*), for example, was sufficiently common in 1592 to receive a mention in Shakespeare's *Henry VI*; a grass closely related to perennial ryegrass, it was prevalent in many arable areas until the late nineteenth century. The grain is toxic, the symptoms typically including impaired vision or blindness (Forsyth, 1979).

Another very characteristic species, also poisonous, was corncockle

Table 1.3: *Weeds and ruderals whose first post-glacial record is from Roman and British (Iron Age) times (c. 500 BC – c. AD 500)*

Arable weeds	Ruderals and grassland weeds
Parsley Piert	Bulbous buttercup
Corncockle	Garlic mustard
Corn marigold	Common mallow
Nipplewort	Hogweed
Scarlet pimpernel	Upright hedge-parsley
Corn gromwell	Oxeye daisy
Spring wild oat	Cat's-ear
Rough poppy	Betony
Fool's parsley	Field woundwort
Fumitory	Hemlock
Runch	Greater celandine
Spurrey	Basil thyme
Black nightshade	Bishopweed
Annual nettle	
Smooth sow-thistle	
Prickly sow-thistle	
Darnel	

(*Agrostemma githago*), which showed a similar decline from abundance in the eighteenth century to virtual extinction in the twentieth. Corn marigold (*Chrysanthemum segetum*) on the other hand, while evidently greatly reduced from former levels, still persists in areas which meet its requirement for a rather light and slightly acid soil. It has the distinction, according to Salisbury, of having been on the receiving end of the 'earliest recorded enactment requiring the destruction of a pernicious weed', in the time of Henry II. Yet another species whose name is suggestive of its former role is the cornflower (*Centaurea cyanus*). Salisbury gives several accounts of this as a troublesome weed between 1640 and the late nineteenth century, including the well-known passage by John Clare describing the 'blue cornbottles . . . troubling the cornfields with their destroying beauty'. Table 1.4 lists a number of species additional to the above which also declined significantly from about 1700 onwards.

This well-documented reduction is a clear indication of an increasing awareness of weed problems. The gradual introduction of the horse-drawn seed-drill after 1730 laid the basis for effective mechanical hoeing between the crop rows (Jethro Tull's 'horse-hoeing husbandry') and appears to have had a marked effect in reducing total weed numbers. A general improvement in soil fertility and the increasing use of lime and marl must also have operated to the detriment of acid-tolerant species such as corn marigold, while drainage brought about a reduction in marsh cudweed, mousetail and others (Salisbury, 1961).

Table 1.4: *Weeds which declined in abundance during the period c. 1700 to c. 1900*

Pheasant's-eye
Larkspur
Night-flowering catchfly
Lesser snapdragon
Thorowax
Field cow-wheat (Poverty-weed)
Starthistles
Spreading hedge-parsley
Rye-brome

Equally important was the incorporation of new and more elaborate rotations, which by varied cropping and cultivations denied any one weed or group of weeds consistently favourable conditions. The new row crops especially came to have a vital cleaning role, to be set against the 'fouling' crops of beans and cereals. These systems and their accompanying technology formed the backbone of weed control until the mid-twentieth century, when the increasing scarcity and cost of labour led to increasing dependence on the newly-developed chemical weedkillers.

Another major factor in the decline of certain weed species was the development of efficient seed-cleaning machinery from the first half of the nineteenth century onwards. Until that time seed-cleaning was no more than an extension of the ancient device of 'winnowing': tossing the crop seed – cereal grains in the main – into the air and allowing the wind to blow off the lighter weed seeds and chaff. Early developments in this area are well described by Harvey (1980) for England and Wales and by Fenton (1976) for Scotland.

In both cases the actual separation or threshing of the grain was carried out by beating the sheaves with a long-handled wooden or leather flail. Harvey refers to the winnowing of the threshed grain in the draught from the open door of the threshing barn, following which it was thrown across the threshing floor by a casting shovel or other device, so separating the 'tailings' or lighter portion from the 'headcorn' containing most of the crop seeds and the heavier weed seeds. Ernle (1961) describes a more elaborate procedure, in which a hand-driven wheel with sacking attached was used to winnow the grain, which was then allowed to run down across a sieve or riddle before finally being thrown. In either case the headings were often winnowed again or else riddled by hand, using a varying mesh size depending on the crop concerned – approximately 4mm. for wheat, 5mm. for barley and 7mm. for oats.

After a number of false starts, what is generally recognised as the first really efficient threshing machine was built by Andrew Meikle in 1786. The idea caught on quickly and both barn and mobile machines were soon widely in operation, with horses or sometimes water providing the motive power. Initially, separate powered winnowers were used, but later these were incorporated in the machines themselves. By the middle of the last century steam-driven mills were taking over and the whole process latterly culminated in the traction-engine-powered travelling mill, the arrival of which on the farm is quoted by many earlier writers as being one of the great events of the farming year.

1.1.3 INTRODUCTION AND SPREAD OF WEEDS, 1700–1900

All the movement was not in one direction however, and the above period also saw increases in the incidence both of some established weeds and also of a fresh crop of introductions, some of which are continuing to spread at the present time.

Among native species, rosebay willowherb (*Chamaenerion angustifolium*) and bracken (*Pteridium aquilinum*) may be taken as examples. The former, now perhaps the commonest colonising plant of waste places, derelict sites and cleared areas generally, was apparently quite rare until about a hundred years ago. The development of industry on a large scale, together with railways and similar habitats, may have provided it with the type of open, disturbed environment which it evidently requires, although fortunately it appears unable to compete successfully under arable conditions. The increase in bracken, which took place mainly during the nineteenth and early twentieth centuries, also had marked socio-economic overtones, including the replacement of cattle by sheep in many upland areas, the abandonment of marginal fields and the cessation of its use as litter.

Two highly invasive alien weeds are also worth a mention here. Following its introduction in 1753, rhododendron (*R. ponticum*) is now thoroughly naturalised on sandy or peaty soils in the north and west especially, often to the exclusion of other species. A later (1825) arrival, Japanese bamboo or knotweed (*Polygonum cuspidatum*), has also become widely distributed, especially along riverbanks or near the sea, again to the detriment of native vegetation. Other examples, in order of introduction, include the following:

1666: Sticky groundsel (*Senecio viscosus*). First recorded right at the start of this modern period, it is only in the last fifty to sixty years that this small composite has really extended its range, possibly in association with the development of modern road transport.

1770: Oxford ragwort (*S. squalidus*). In the case of this closely related species it seems to be generally accepted that transport, this time by rail, has again been a crucial factor.

1809: Hoary cress (*Cardaria draba*), otherwise known as peppercress or hoary pepperwort (from a former use of the seeds) or Thanet cress (from the locality into which it was introduced). A perennial, it has rather inefficient seed dispersal, but an extensive system of creeping roots, capable of producing daughter shoots anywhere along their length. This has given rise to a rather slow and discontinuous spread over the country as a whole – it is still confined as a weed problem mainly to south and central England – but rapid development from each initial 'infection focus'.

1825: Common field speedwell (*Veronica persica*). Also known as Buxbaum's speedwell. Introduced possibly on more than one occasion, it has spread rapidly through England and southern Scotland and has now superseded the indigenous speedwells in many arable areas. Another Asiatic speedwell, *V. filiformis*, the slender speedwell, first introduced in 1927, has also become widespread, in this case as a pernicious weed of lawns and rock gardens.

1842: Canadian waterweed (*Elodea canadensis*). Like several other water-plants this N. American species showed an extremely rapid spread in inland waterways after its introduction, followed by consolidation at a more manageable level.

1850: Small-flowered balsam (*Impatiens parviflora*). The most important of three species of balsam introduced in the early eighteenth century. A difficult weed of shaded parts of gardens on richer soils.

1860: Gallant soldier (*Galinsoga parviflora*). Otherwise known as Kew-weed, following its escape from the Botanic Gardens, this has had a well-documented spread through the southern counties of England. Its common name is a corruption of the Latin.

1871: Pineappleweed (*Matricaria matricarioides*). A N. American plant, which spread particularly in the period 1900–1925. It is typically found around field gateways, farmyards and farm roads and tracks, and is evidently tolerant of soil compaction and surface traffic.

Mid-nineteenth century: Comfrey (*Symphytum* spp.). Three species of comfrey were introduced in the mid-nineteenth century and have become established as roadside plants, occasionally invading arable fields. As 'Russian comfrey' they have more than once received attention as a possible stock feed.

Mid-nineteenth century: Treacle mustard (*Erysimum cheiranthoides*). First recorded in medieval times, its ultimate spread as a weed of gardens and horticultural crops especially appears to have resulted from reintroduction in the last century. Continuing to spread.

Early twentieth century: Amsinckia (*Amsinckia intermedia*). A N. American species related to the above. Established and still spreading as an arable weed on lighter soils, in the south-east especially.

1917: Winter wild oats (*Avena ludoviciana*). Again from America, and of all these plants much the most important, as a major weed of winter cereals.

Certain points emerge from this short review which we may take as being valid in the case of earlier, undocumented introductions. Firstly there is the concentration of new arrivals in the middle of the nineteenth century, a time when agricultural trade and the movement of grain and seeds were expanding rapidly, between Europe and N. America in particular. We can see parallels here with previous periods when comparable large-scale adjustments in trade were taking place, for example in Roman times.

It is notable too that in many of these alien species a sometimes lengthy lag-phase preceded eventual expansion. This may reflect simply the later establishment of some factor or factors especially favourable to the species concerned (e.g. open ground in the case of rosebay willowherb), or it may involve a more subtle mechanism, related to the somewhat elusive concept of 'infection pressure', which implies that the population of a given species must attain a certain threshold level before it can break out and achieve a major extension of its range.

It is apparent also that many of the more recent introductions are still spreading at present, sometimes at the expense of their native relatives. The hypothesis that introduced species are less subject to attack by pest and disease organisms is no doubt relevant here, but it is clear that genetic/ecological considerations are also involved.

Baker (1974) provides a review and an extensive bibliography of this topic and suggests, *inter alia*, that native or other long-established species, having evolved the ability to survive in a given environmental/biological context, may have lost some of their genetic variability. They may, therefore, be at a disadvantage against an incoming species, particularly if the latter possesses what Baker refers to as a 'general purpose' genotype, enabling it to compete over a wide range of circumstances. The incomer's competitive ability may be further enhanced if it also features self-fertilisation and/or some form of apomixis (reproduction without fertilisation) including efficient vegetative propagation, enabling rapid multiplication of genetically similar individuals.

1.2 Present status and distribution

1.2.1 IMPROVEMENTS IN SEED-CLEANING

We entered this century, therefore, with a weed flora of long standing, which closely reflected developments in agriculture during earlier periods. This process has continued through, for example, the introduction of increasingly sophisticated seed-cleaning techniques, including magnetic separation of seeds following coating with iron filings, electrostatic separation, and colour-separation by means of photoelectric cells. Modern mobile and static cleaners, often employing a variety of separation mechanisms, are now of course an indispensable feature of the drive towards cleaner crop seed.

The need for administrative support was also increasingly recognised, culminating in the setting up of Official Seed Testing Stations in Scotland in 1914 and in England and Wales in 1917, and in the Weeds Act of 1920, which required a declaration by a seller of seed as to its purity and germination. Even then there was still considerable scope for the transference of weed seeds, which led to the development of voluntary arrangements such as the British Cereals Seeds Scheme, designed to offer growers improved standards over the legal requirement. More recently the 1974 Plant Seeds and Varieties Act has altered the situation to one in which seed which does not reach prescribed levels of purity etc. may not legally be offered for sale. This approach, which incorporates both a legal minimum and a higher voluntary standard (HVS), provides the best hope yet of successfully breaking this vital link in the spread of weeds, but could be undermined by lack of attention to the cleanliness of farm-sown seed, especially in cereals. In tests at the official Seed Testing Station, Cambridge, between 1978 and 1981, many more weed seeds were found in farmers' own grain than in commercial first and second generation samples. In both cases grass seeds and other cereals constituted the main impurities, but broad-leaved species also occurred, notably large-seeded types such as black-bindweed and cleavers, which are difficult to clean from cereal grain (Tonkin, 1982). A comprehensive account of weed problems and control measures in seed crops generally is given in Roberts (1982).

1.2.2 THE INTRODUCTION OF HERBICIDES

An even more far-reaching event was the introduction in the 1940s of the synthetic growth-regulator or 'hormone' weedkillers to control broad-leaved weeds in cereals. The rapid increase in the use of MCPA, 2,4-D and related chemicals from the early 1950s onwards greatly reduced and in some cases virtually eliminated many remaining survivors of the traditional cornfield flora, including pennycress, poppy, corn buttercup and others (Fryer and Chancellor, 1970).

Concurrently with this, however, the principle that 'nature abhors a vacuum' came into play and these susceptible species were soon replaced by more resistant types such as cleavers, mayweeds, chickweed, and the 'polygonums' – redshank, knotgrass and black bindweed. Despite the later introduction of more potent herbicides and mixtures, this group still dominates the weed scene at present, although a few MCPA-susceptible species such as charlock continue to flourish, reflecting perhaps a high survival rate of buried viable seed. The picture is completed, so far as broad-leaved weeds are concerned, by certain especially resistant species, including speedwells and corn marigold; although controls are now available, these can still be important in areas where husbandry and soil factors combine to favour them.

Changes have also taken place in the much smaller group of grass weeds, related to the continuing upward trend in cereal-growing. These have involved increases in blackgrass and latterly sterile and other brome grasses, while perennial species such as rough-stalked meadow-grass and ryegrass can also cause problems.

1.2.3 REGIONAL DISTRIBUTION OF WEEDS

The current situation, therefore, is one of dominance by a relatively small group of major species, together with others which may be important in particular areas or crops. Detailed information on weed distribution and abundance derives mainly from the long-term obser-vations of farmers, advisers and research workers, together with a small number of more or less objective surveys. O'Leary (1973), for example, on the basis of a nationwide survey of broad-leaved weeds in cereals, identified some twenty species of national importance, each infesting 10% or more of the acreage of spring or winter cereals, or both (Table 5.1, p. 89).

The distinctive nature of the weed flora was emphasised by Way (1972), who found little overlap between species of cultivated ground and those of adjacent habitats such as woodland and hedgerows. This point is also brought out through the linking of British and Continental weed communities in the context of the overall classification of European vegetation. The system involved stresses the role of ecological factors and of typical or 'character' species in establishing links between related vegetation types. On this basis Haflinger and Brun-Hool (1971) recognised 31 weed associations in Europe, many of which are only poorly represented in Britain, because of our restricted range of species.

Nevertheless, a number of regional variations in weed distribution have been identified. O'Leary's survey, for example, showed a prevalence of autumn-germinating species such as cleavers, mayweeds and speedwells in the winter-cereal areas of central, southern and eastern England, while elsewhere in England and Wales and in Scotland redshank and fumitory were relatively abundant. Scotland and N. Ireland also had high levels of hemp-nettle and the latter especially much acid-tolerant spurrey and corn marigold. Scragg

Table 1.5: *The regional distribution of broad-leaved weeds in the United Kingdom*

	% of total hectarage infested	North	South	East	West
Shepherd's purse	5	XX	XX	XX	XX
Hemp-nettles	5	XXX	X	XX	X
Cleavers	16	X	XX	XXX	X
Deadnettles	5	XX	XX	XX	XXX
Mayweeds	18	XX	XX	XXX	XX
Knotgrass	8	XX	XXX	XXX	XX
Black bindweed	6	XX	XXX	XXX	XX
Pale persicaria and Redshank	10	XX	XXX	XXX	XX
Charlock	9	XX	XXX	XX	XX
Chickweed	33	XXX	XXX	XXX	XXX
Annual nettle	6	X	X	XX	X
Speedwells	5	X	XX	XXX	XX
Field pansy	5	X	XXX	XX	XX

Key to relative importance:
X – locally occuring, seldom a problem
XX – widely occurring, sometimes a problem
XXX – commonly occurring, often a problem.

(After Makepeace, 1982, a)

(1974) reviewed this aspect in detail and pointed to the existence of a specifically 'northern' weed flora in north-east Scotland, for example, reflecting the generally higher organic matter, moisture content and acidity of the soils and the effect of lower autumn and winter temperatures on germination and survival.

The ten years or so since these observations were made have seen an increasing divergence between the main arable regions of England and elsewhere (Makepeace, 1982,a). In that time the area of winter cereals has doubled, reinforcing regional distribution patterns (Table 1.5). A major factor in this has been the trend to earlier sowing and the increased use of tine cultivations and direct-drilling, one result of which has been an increase in annual grasses and some perennial weeds, at the expense of annual dicots (Cussans *et al.*, 1979). This confirms fears expressed as to the effects of intensified arable cropping and, in particular, an increased dependence on herbicides, in creating conditions where new weeds can arise with alarming rapidity (Cussans, 1980) and reminds us once again of the essentially volatile nature of the problem.

Further Reading

FROUD-WILLIAMS, R. J., CHANCELLOR, R. J. and DRENNAN, D. S. H. (1981), 'Potential changes in weed floras associated with reduced-cultivation systems for cereal production in temperate areas', *Weed Research, 21*, pp. 99–109.
> A full review of the effect of reduced-cultivation systems on the weed flora, actual and predicted, with reference also to the build up of resistance and to weed dispersal strategies.

FRYER, J. D. and CHANCELLOR, R. J. (1970), 'Herbicides and our changing weeds', in PERRING, F. (ed.) *The Flora of a Changing Britain*, Report No. 11, Botanical Society of the British Isles, London.
> Factual account of changes in the weed-flora associated with the deployment of synthetic herbicides post-1945.

GODWIN, H. (1960), 'The History of Weeds in Britain', in HARPER, J. L. (ed.), *The Biology of Weeds*, Blackwell Scientific Publications, Oxford.
> Condensed, readable version of the author's major (1956) work. Much else of relevance in this volume.

SALISBURY, E. (1961), *Weeds and Aliens*, Collins, London.
> The definitive treatment of weed origins and distribution in Britain. Profusely illustrated.

SCRAGG, E. B. (1974), 'Regional weed problems – dicotyledonous weeds in tillage crops', *Weed Control in the Northern Environment*, Br. Crop Prot. Coun. Monogr. No. 10, pp. 19-32.
> Concise coverage of weed distribution and underlying soil, climate and husbandry factors.

TONKIN, J. H. B. (1982), 'The presence of seed impurities in samples of cereal seed tested at the Official Seed Testing Station, Cambridge, in the period 1978–1981', *Aspects of Applied Biology 1*, 1982, Broad-leaved Weeds and their Control in Cereals, pp. 163–71.
> Useful account of the position re seed impurities, after several years of 'compulsory' quality standards. Also covers grass weeds.

2 The Biology of Weeds

Before considering in detail the biology of weed plants, we should try to define exactly what we mean by the term 'weed'.

Some would argue that there is no such thing as a weed and that all plants are of equal importance and should be treated accordingly. However, farmers and growers cannot afford to be so democratic. They tend to divide plants into two categories: namely, the useful plants or crops, from which man and his animals derive benefit, and the troublesome plants or weeds which interfere with agricultural/horticultural practices.

In the broadest sense, a weed is a plant growing in any situation where it is considered undesirable. By definition, therefore, a plant need not always be a weed. Indeed, potatoes grown as a crop one year can present a serious weed problem in a variety of agricultural and horticultural crops in succeeding years. This, however, is an extreme example, and the majority of plants classified as weeds, and commonly recognised as such, possess certain characteristics which account for their success, and which make their eradication extremely difficult to achieve.

Weed species are, in general, extremely competitive, aggressive colonisers and have the ability to adapt to various environments. Arguably, however, their most important attributes are their efficient reproductive systems, coupled with the possession of mechanisms which enable them to survive in temporarily unsuitable environments.

2.1 Effects of weeds

Weed plants fall into two main categories, broad-leaved weeds (dicotyledons) and grasses, rushes, sedges and bulb-plants (monocotyledons). Both types produce adverse effects in the crop situation and in many other man-made environments.

2.1.1 THE DETRIMENTAL EFFECTS OF WEEDS ON CROPS

These include the following:

a) *Competition for light, water and nutrients, both during establishment and in the established crop*

Weed competition during crop establishment is particularly important. In newly planted apples, for example, there is evidence that shoot production is closely correlated with weed growth. In trials involving the variety Cox's Orange Pippin, competition from a 50% ground cover of weeds around young trees was found to produce severe reductions in shoot length in the first year of growth, compared with weed-free plots. This reduction persisted in the following two years, even when the plots were maintained in a weed-free condition: it was not remedied by application of extra nitrogen, and was attributed mainly to moisture stress (White and Holloway, 1967).

In raspberries too, heavy weed cover in the late spring/early summer in the year of planting may produce shorter canes and severe reductions in the number of canes, resulting in over 50% less fruit being picked in the second year.

Failure to control weed growth in established crops may cause reduction in crop growth and yield. For example, removal of couchgrass (*Agropyron repens*) in spring resulted in increases of 30–40% in cane production in raspberries (Lawson and Rubens, 1970). Similarly, the yield of strawberries can be increased by judicious and timely control of weeds.

b) Interference with harvesting operations, handling and quality

In the days of binders, weeds hindered harvesting little, except by causing discomfort to the workers. However, combine harvesting has become an increasingly complex and carefully balanced operation, where the presence of weeds can lead to problems. For instance, weeds alter the ratio of total matter to grain going into the combine. They may also, if green, affect the moisture content and the consistency of the bulk sample being processed. If the crop is laid and contains green weeds, there may be difficulty in ensuring collection on the combine table and regular feeding of the drum.

In orchards and soft fruit establishments, weeds also interfere with harvesting. Species such as nettles (*Urtica* sp.) and thistles (*Cirsium* sp.) may discourage pickers, thereby reducing harvested yields. Again, unpleasant conditions underfoot, or inability to pick quickly because of weeds, may divert pickers and pick-your-own customers to pleasanter, more rewarding, plantations.

The time required for lifting nursery stock will also be greatly increased if roots of perennial weeds have to be removed from the roots of the nursery stock during lifting and before sale.

In addition to affecting crop yield, weeds can also affect crop quality. If crops grown for seed contain weed seeds of like size, shape and density, this may prevent the crop from being sold as a seed crop. Also, some weeds such as darnel (*Lolium temultentum*) and corncockle (*Agrostemma githago*) have, in the past, caused poisoning to man when ground into flour. However, this is now prevented by seed cleaning.

Weeds such as ragwort (*Senecio jacobaea*) and bracken (*Pteridium aquilinum*) are poisonous to stock and may cause severe losses. All classes of farm livestock are susceptible to poisoning by these plants, cattle and horses much more so than sheep. In the case of ragwort, palatability is increased both by herbicide treatment and wilting, but by far the greatest danger of ragwort poisoning arises from livestock eating contaminated silage. None of the toxicity of this plant is lost in silage, and cattle cannot discriminate between the ragwort and other material.

c) Weeds acting as alternate hosts for pests and diseases, giving shelter to vermin, or diverting pollinating insects

Several weed species act as hosts for a range of fungal, virus and nematode-borne diseases of crops. In cereals, couchgrass (*Agropyron repens*) has been linked with take-all and blackgrass (*Alopecurus myosuroides*) with ergot, both fungal diseases of wheat.

Volunteer cereal plants can also be a source of disease in a following cereal crop. Yellow rust of wheat, and mildew and brown rust of spring barley have been shown to be associated with volunteer wheat and barley plants respectively.

The seeds of some common weeds including common chickweed (*Stellaria media*) and corn spurrey (*Spergula arvensis*) have the capacity to transmit and spread nematode-borne diseases. This can occur in two ways. Either the nematodes can become infective from seedlings produced from contaminated seeds, or the disease organism can be carried from one crop to another in the infected weed seeds themselves.

Weeds can also create problems for the farmer/grower by providing shelter for vermin. Severe injury to apple tree trunks by mice can result from dense weed growth around the base of the trees. Conversely, a weed-free strip along fruit tree rows apparently affords protection against winter damage by voles.

An unusual source of weed competition with fruit trees is illustrated by the report (Free, 1968) that, at blossom time, the pollen of dandelion (*Taraxacum officinale*) may be more alluring to bees than the pollen of either apple or plum, so that removal of this weed from orchards could enhance the pollinating efficiency of honey-bee colonies.

2.1.2 NON-CROP SITUATIONS

In non-crop situations weed control may also be desirable. Aquatic weeds for example can create problems when their growth becomes excessive, by impeding flow, blocking pumps and sluices, and interfering with the recreational use of water. In sports fields and other areas of amenity turf, weed control is beneficial for aesthetic reasons, and because certain weeds can interfere with the activities for which the playing surfaces were designed.

In industrial conurbations too, especially where there is a fire risk, on railway tracks and on tennis courts, complete eradication of vegetation is desirable. Partial control of vegetation may also be required, as in the case of motorway verges and riverbanks.

2.2 Types of weeds

A major factor influencing the success of weeds is their ability to reproduce readily. A knowledge of weed life-cycles and, in particular, their reproductive systems is important in planning any programme for their control. It is helpful, therefore, to group certain weeds together according to their life-cycles, but it should be emphasised that these are only loose groupings and there is a degree of overlap between the different categories.

In the first group, plants complete their life-cycle in one season. The seed germinates, usually in the spring or summer, and the plant develops and finally produces its own seed, all within the space of a year. Such weeds are termed ANNUALS and many of the most common agricultural/horticultural weeds belong to this group, including fat hen (*Chenopodium album*), redshank (*Polygonum persicaria*), annual nettle (*Urtica urens*), charlock (*Sinapis arvensis*) and red deadnettle (*Lamium purpureum*). Groundsel (*Senecio vulgaris*) produces seeds which germinate quickly to complete two or three generations in a single year. This type of plant is sometimes referred to as an EPHEMERAL.

Some weeds take two years to complete their life-cycle. They develop vegetatively in the first year and produce seeds in the second, and are therefore called BIENNIALS.

ANNUALS and BIENNIALS are characterised by the fact that they die after producing one lot of seed or fruit. An example of a biennial weed is ragwort (*Senecio jacobaea*), which only rarely flowers in the first year. In the second year the plant appears in the form of a flat rosette of leaves close to ground level, and during the summer the stem elongates to bear the flowers and later the seeds. Ragwort also serves to illustrate the overlap between the different categories of weeds, as it may also be considered to belong to the third group, the PERENNIALS. These plants go on living year after year, sometimes producing seeds every season and sometimes only once in several seasons. While the aerial shoots of such plants often die back each year after flowering, the underground parts remain alive and send up new flowering shoots in the following season. This group contains many important weeds including couchgrass, (*Agropyron repens*), dock, (*Rumex* sp.), bishopweed (*Aegopodium podagraria*), and bracken (*Pteridium aquilinum*).

2.3 Reproduction and Spread of Weeds

Weeds reproduce by seeds, vegetative methods, or both, depending on the species and the growing conditions.

2.3.1 REPRODUCTION BY SEED

Seeds fulfil two major purposes in the life of plants, namely the dispersal of new individuals for the colonisation of new habitats, and the maintenance of the species under unfavourable external environmental conditions.

Although most common weeds generally produce seed at some stage during their lifetime, there are a few exceptions including field horsetail (*Equisetum arvense*) and bracken (*Pteridium aquilinum*), which are non-flowering plants and produce spores instead of seeds.

The fact that weeds produce vast numbers of seeds (including fruits) undoubtedly gives them an advantage in any competitive environment. Estimates of seed production per plant from some of the common weeds include groundsel 1000–1200 (fruits), shepherd's purse (*Capsella bursa-pastoris*) 3500–4000, mayweed (*Tripleurospermum maritimum* spp. *inodorum*) 15000–19000 (fruits), common poppy (*Papaver rhoeas*) 14000–19500, and rosebay willowherb (*Chamaenerion angustifolium*) up to 80000. The significance of these seed numbers may be demonstrated by the fact that the average number of seeds produced by a wheat plant would be in the region of 90–100.

Generally, crop plants tend to produce all their seeds in one burst. While this is also true of some weeds like charlock (*Sinapis arvensis*), others like annual nettle (*Urtica urens*) begin producing seed while the plants are still quite small, and continue to do so throughout the season.

Although seed production by individual plants is important, from a practical viewpoint the number of seeds produced by the entire weed population of a particular area is of greater significance. Vast numbers of seeds from many species will be shed and may either germinate immediately or lie in the soil, viable but not germinating, for varying periods of time. In either event they present a tremendous potential for competition to both existing and succeeding crops.

Experiments on populations of weed seeds of arable soils have shown that there may be up to 86,000 viable seeds/m². While the number of weed seeds in the soil tends to be smaller when a crop is present, the extent of the problem is highlighted by the fact that, on average, the seed rates for spring barley and carrots are approximately $400/m^2$ and $130/m^2$ respectively.

It is clear, therefore, that populations of weed seeds in the soil pose a threat to the growth and success of any proposed crop.

In general weed seed populations are increasing annually, for a variety of reasons. Seeds may be sown with the crop, or imported from other areas by the wind, animals, or by man himself. However, in circumstances where no more seeds are added, the population normally decreases by between 20% and 50% per annum. This decrease depends on several factors including germination, seed mortality and decomposition by micro-organisms, and varies with species, cultivation and the characteristics of the soil.

2.3.2 GERMINATION OF WEED SEEDS

Weed seeds, like other seeds, require certain conditions before they will germinate.

Water is necessary to replace that lost by the embryo and other tissues during seed ripening, for the activation of enzymes, and for transport to the embryo of the mobilised food reserves.

Oxygen is required for aerobic respiration, which provides energy for the growth of the embryo.

A suitable temperature is also necessary. This varies with different seeds. Increasing the temperature from minimum to optimum level also increases the germination rate.

The effect of light on germination is variable, the majority of seeds germinating equally well in light or darkness. However, some seeds germinate successfully only in the light, while seeds of other species will not germinate unless in the dark.

2.3.3 DORMANCY

In the developing seed the embryo ceases to grow as the seed ripens and all parts of the seed lose water. By the time the seed is fully ripe all the physiological activities of the embryonic and other living cells are reduced to a minimum.

In certain cases (e.g. willow), if conditions are favourable, the seed is able to germinate immediately it leaves the parent plant. In the majority of plants, however, the ripe seed, dehydrated and physiologically inactive, will not germinate immediately, even when the environmental conditions are suitable. The seed is then said to be in a state of DORMANCY, when growth is temporarily suspended, and all physiological processes appear to be almost at a standstill.

The failure of dormant seeds to germinate even under favourable conditions may arise from a number of factors, which may either be imposed by the embryo coverings or may be a property of the embryo itself. This PRIMARY or INNATE dormancy is genetically controlled, and is independent of environmental factors.

When dormancy is the result of the enclosing seed or fruit coat, it may be related to one of the following properties:

a) Impermeability of the coats to water

The seeds of certain families, such as *Leguminoseae* (e.g. clover) and *Chenopodiaceae* (e.g. fat hen) possess testas which are impermeable to water, and are liable to lie dormant in the soil for considerable periods before germination occurs. Such seeds are termed 'hard' seeds, and water intake is prevented by the thick-walled cells of the testa which is surrounded by a hard waxy layer. Rupture of the testa at a later date immediately allows water to enter the seed, after which germination soon begins. Under natural conditions these seeds lie in the ground without germinating for varying lengths of time before being released from dormancy by mechanical damage, insect damage or microbial decomposition of the testa.

b) Impermeability of the coats to gases

The fruit of cocklebur, a common weed in the United States, contains two seeds, one dormant and one non-dormant. Removal of the testa from the dormant seed permits germination to proceed. It has been demonstrated that portions of excised testa do not allow the diffusion of oxygen, provided they are undamaged. For many other types of seed, including grasses, an increase in oxygen concentration in the surrounding air and damage to the seed coats allow an increase in the rates of respiration which often results in germination.

c) Mechanical resistance to embryo growth

The seeds of the rose, in common with many other plants, are enclosed in hard stony layers which require strong forces to break them down. This, apparently, is only part of the story since there is evidence that dormancy-inducing hormones are present in rose seeds and, indeed, the unchilled embryo may not start to grow even when the coat is removed.

In addition, dormancy may be induced by the embryo itself.

1. Immaturity of the embryo   Seeds of some species, for example lesser celandine (*Ranunculus ficaria*), possess embryos which are not completely developed, and germination cannot proceed until this has been achieved. This normally occurs in the seed during autumn and winter, and the seed finally germinates in the spring.

2. After-ripening in dry storage   The seeds of many species will not germinate if sown immediately after harvesting, even though the

embryo is fully developed. They require a period of drying, after which they gradually lose their dormancy and will germinate given favourable conditions. This characteristic of after-ripening in dry storage is found in several cereals, for example barley, wheat and oats. It is not known what causes this type of dormancy or how it is broken, but it would appear that it is not under metabolic control, since it occurs even in dry seed where metabolism is minimal.

3. Requirement for light   In numerous species, exposure to light is necessary before germination can proceed, e.g. curled dock (*Rumex crispus*) and great willowherb (*Epilobium hirsutum*). Such seeds are termed 'light-sensitive' seeds. Conversely, germination of some seeds is inhibited by light ('light-hard' seeds). Many plants require this dormancy-breaking mechanism only when the seeds are newly produced, and the need for light often disappears during dry storage when, presumably, certain changes take place which remove the requirement for light. Light-sensitive seeds will only germinate after they have imbibed water, and subjecting air-dry seeds to light produces no response.

4. Dormancy removed by chilling   It has long been known that the seeds of a number of species require exposure to low temperatures before they will germinate.

This characteristic gave rise to the practice in horticulture whereby seeds are stratified, or placed between layers of sand and kept outside during the winter. Come the spring, these seeds are no longer dormant and germinate readily.

In the past it was thought that this type of dormancy was caused by hard and impermeable seed coats and that freezing temperatures were necessary to break the coats. However, it is now known that temperatures just above freezing (0–5°C) are more effective and, in addition, many seeds which require chilling do not, in fact, possess hard coats.

In some cases, cold-treatment is a prerequisite for germination. In others, while it increases and hastens germination, it is not essential. Cold-treatment is only effective if seeds contain water, and treatment of dry seeds has no effect.

Certain plants produce 'two-year seeds', so called because they do not usually germinate until the second spring after shedding. Some of these seeds possess hard coats in addition to requiring cold-treatment. This means that the embryos cannot imbibe water on being shed so that chilling during the first winter does not break dormancy. However, the activities of soil micro-organisms in the following summer enable the seeds to take up water, as a result of which dormancy is overcome during the second winter and the seeds can germinate in the spring.

Some seeds may exhibit more than one type of dormancy, which means that more than one type of dormancy-breaking treatment is required to enable germination to take place.

Secondary Dormancy

Seeds of certain plants which do not possess PRIMARY dormancy and which would, under normal circumstances, germinate freely, may become dormant in response to an unfavourable environment. This condition is termed SECONDARY dormancy and can be overcome by one or other of the dormancy-breaking mechanisms previously described.

Plant breeding and selection over the years has meant that seeds of the major crop plants are not dormant to any great extent, and man can sow his seed when he considers conditions to be most suitable. In nature, however, the seed itself must possess properties to enable it to judge the most appropriate time for germination.

The seeds of many arable weeds normally germinate soon after being released from the parent plant if conditions are favourable. Under certain circumstances, however, for example if they are buried, they may develop a state of secondary dormancy and will not germinate until freed from the conditions which caused them to become dormant. Such seeds may lie in the soil for varying lengths of time, and will only germinate when they are brought to the surface by ploughing, for example. There are indeed several reports that among weed seeds brought to the surface after being buried for many years under grassland, a large proportion were still capable of germination.

2.3.4 DEPTH OF GERMINATION

Most seeds, including weed seeds, have a maximum depth from which they can germinate and emerge from the soil. This depends both on seed size and food reserves, which must be sufficient to enable the germinating seedling to reach the soil surface and undertake production of its own food by photosynthesis.

The majority of annual weed seeds are to be found in the top 5 cm of the soil (Fig. 2.1), although there are a few which can emerge from greater depths.

Depth of germination must be considered in any system of weed control. If cultivation is to be employed to stimulate germination of weed seeds, the first pass in a series should be the deepest, so that deep-lying seeds are brought to the surface where they will germinate. This is very important because, with the exception of the soil-sterilant chemicals (dazomet and methyl bromide), soil-applied herbicides do not normally control weed seeds which are not in the process of germinating. It is necessary, therefore, to encourage the seeds to germinate, by ploughing for example, so that they can be controlled by herbicide treatment.

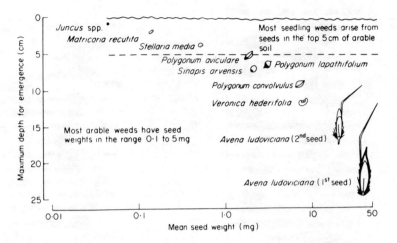

Fig. 2.1: *Relationship between seed size and the maximum depth of germination from which seedling emergence can occur (after Roberts, 1982).*

With certain herbicides, it is recommended by the manufacturer that they should be incorporated into the soil within a certain time after application. This is because they may be broken down by sunlight and thus inactivated. However, there is evidence to suggest that, even in cases where incorporation is not essential, the performance of soil-applied herbicides can be improved by this practice. The reason for this is that incorporation transfers the herbicide to the region where it has the best chance of coming in contact with germinating weeds, and its performance is therefore less dependent on rainfall. Hence, depth of germination is an integral factor in determining how deep soil-applied herbicides should be incorporated.

2.3.5 PERIODICITY OF GERMINATION

While the odd seedling of many weed species may emerge throughout the year, the vast majority of seeds of any one species germinate at one particular period in the year (Fig. 2.2). This is termed PERIODICITY OF GERMINATION, and is important for several reasons. It means that there is a close association between certain weeds and crops. For example, ivy-leaved speedwell (*Veronica hederifolia*) germinates mainly in the autumn and is therefore particularly troublesome in winter cereals. It is also important in relation to the use of soil-applied herbicides when,

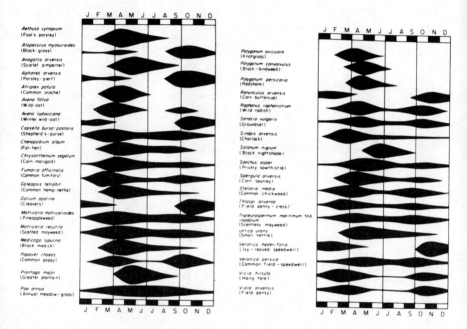

Fig. 2.2: *The main germination periods for some common annual weeds of arable land (after Roberts, 1982).*

if the period of maximum germination of a particular species is known, the timing of the application of an appropriate soil-applied herbicide or herbicide mixture can be geared to coincide with this. It is advisable, therefore, for a farmer or grower to keep a record from year to year of the main weed species present in each of his fields.

2.3.6 SPREAD OF WEED SEEDS

While it is generally agreed that man is the most successful agent in the dispersal of weed seeds, natural agencies and appropriate modifications of the plant are also important, as follows:

WIND Structural modifications of many seeds aid their dispersal by wind. The parachute-like pappus of the compositae (e.g. dandelion) enables the seeds to be spread over relatively long distances.

A number of weeds have active dispersal mechanisms and when they are ripe, they violently eject their seeds away from the parent plant. One of these is hairy bitter-cress (*Cardamine hirsuta*), a common horticultural weed. The problem with this plant is that each time it is disturbed, it releases its seed and this makes it particularly difficult to control, especially in the garden where hand-weeding is generally employed. Where possible, this weed should be removed before it flowers.

WATER Many kinds of weed seeds, even those without special modifications, are carried by water. In particular, docks and sedges have fruits which float easily and are readily dispersed in water.

ANIMALS AND BIRDS Adaptations enabling seeds to be distributed attached to animals are common, e.g. cleavers (*Galium aparine*). The ability to germinate after passage through the alimentary tract may also be important and this has been found in shepherd's purse (*Capsella bursa-pastoris*), common fumitory (*Fumaria officinalis*), annual meadow-grass (*Poa annua*), and common chickweed (*Stellaria media*) amongst many others. Indeed, cases of germination actually improving after passage through an animal have been recorded, but this is probably not common.

MAN Humans are very important in the dispersal of weeds. Indeed, the transfer of weed species over major geographical barriers such as oceans is almost always a result of human agencies.

Weeds may be spread through human agency in a number of ways. For instance, crop contamination by weed seeds is an important means of dispersal. Despite the fact that crop seeds are subjected to very thorough cleaning processes, it is inevitable that weed seeds will be sown along with the crop. Furthermore, weed seeds introduced in this fashion are more than likely to be strong competitors with the crop and hence will provide a potentially serious problem.

In order to avoid this sort of situation the UK, in common with many others, operates strict regulations governing the presence of weed seeds in commercial seed samples, and there are minimum standards of purity for each seed crop below which seed may not be marketed.

Whilst not of such major importance, weed infestation has been reported from animal foodstuffs and bedding. Also, movement of soil either in small amounts on agricultural machinery, or in relatively large amounts as in motorway construction, can account for the dispersal of weed seeds. Moreover, dispersal of weed seeds in straw and other packaging material must be commonplace, as is their spread by the transfer of container-grown nursery stock from one area to another.

Furthermore, certain species brought into this country as garden plants have now become a problem. A striking example is Japanese knotgrass (*Polygonum cuspidatum*) which was introduced in 1825. This plant is now becoming a weed of considerable importance and until the development of glyphosate was expensive and difficult to control.

2.3.7 VEGETATIVE REPRODUCTION

Despite the fact that they do not normally produce large numbers of seeds, perennial weeds can be troublesome and difficult to control.

Couchgrass (*Agropyron repens*) often produces seeds but not necessarily every year, and they have little dormancy. Other perennial weeds have built-in impediments to seed production. For example, creeping thistle (*Cirsium arvense*) is dioecious (male and female flowers on different plants) and only produces seed if the two sexes are less than 100ft apart. It would appear that this does not occur very often, for seed is relatively uncommon. Thus the thistledown that blows about in the summer is of little significance since when the plant does produce seed, it separates from the pappus at an early stage and the seed is normally left behind on the head.

Despite deficiencies in seed production many perennial weeds possess extremely effective methods of vegetative reproduction, i.e. they produce new individuals from their vegetative system. The parts which regenerate vary from species to species. Some have creeping stems which are below the ground (RHIZOMES), as in Japanese knotgrass (*Polygonum cuspidatum*) and bishopweed (*Aegopodium podagraria*), or above ground (STOLONS), as in creeping buttercup (*Ranunculus repens*). Some species such as perennial sow thistle (*Sonchus arvensis*) and field bindweed (*Convolvulus arvensis*) have creeping roots, while others (often biennials) have swollen, non-creeping tap roots, for example dandelion (*Taraxacum officinale*) and spear thistle (*Cirsium vulgare*). Regeneration by vegetative means may occur naturally by rotting and separation of the vegetative parts or, alternatively, it may be encouraged by cultivation, digging and hoeing. All weeds with creeping parts have great potential for regeneration and even small pieces of vegetative organs possess this ability. Dandelion, for example, can regenerate from all levels of its tap root, while curled dock (*Rumex crispus*) if spudded out to a depth of three to four inches will not regenerate as the lower levels of the root are incapable of producing new shoots.

Some weeds like lesser celandine (*Ranunculus ficaria*) produce bulbils which can remain dormant in the soil for long periods, and are extremely difficult to control.

Tubers are produced by field horsetail (*Equisetum arvense*). This is a difficult weed to control, the problem being that there is apparently very little uptake and movement of herbicide from the shoots to the roots and tubers. Therefore, while a reasonable degree of control of the aerial shoots can be achieved, there is virtually no effect on the tubers, which will produce new plants under favourable conditions.

Canadian water-weed (*Elodea canadensis*) in common with certain other water weeds, seldom produces seed in this country. Instead it produces vegetative buds (turions) which separate from the plant in autumn, overwinter in the mud at the bottom of a pond, and produce new plants in the spring when the temperature rises.

The advantage of vegetative reproduction is that each new plant so produced begins life with a greater food supply than would be possible had it grown from a seed. There are, however, several disadvantages. Each daughter plant is genetically identical to the parent, which does not allow for selection and adaptation to changing or difficult environments. In addition, with vegetative reproduction there are no long-distance dispersal mechanisms. While some plants, e.g. creeping thistle, under ideal conditions can produce creeping roots up to 40ft long in a single year, this bears little comparison to the tremendous distances which can be covered by seeds.

With few exceptions, perennial weeds are not normally able to survive continuous cultivation. However, under modern husbandry systems, particularly in horticultural situations, which rely on minimum cultivation linked to the use of soil-applied herbicides, perennial weeds are of increasing importance. With their underground food reserves they are generally unaffected by soil-applied herbicides, and spot treatment with foliage-applied translocated herbicides will be required for their control.

Further Reading

FLETCHER, W. W. (1974), *The Pest War*, Basil Blackwell, Oxford.
    A good basic introduction to weed biology is to be found in sections throughout this text.
HILL, T. H. (1977), *The Biology of Weeds*, Edward Arnold Ltd, London.
    A concise, easily read and understood account of weed biology, which includes some practical exercises.
KLINGMAN, G. C. and ASHTON, F. M. (1975), *Weed Science: Principles and Practice*, John Wiley and Sons, New York, London and Toronto.
    A detailed account of the biology of weeds, with a definite American bias in the crops sections.
ROBERTS, H. A. (ed.) (1982), 'The biology of weeds', *Weed Control Handbook*, 7th edn., Blackwell Scientific Publications, Oxford, pp. 1–36.
    A detailed account of all aspects of weed biology.
STEPHENS, R. J. (1982), *Theory and Practice of Weed Control*, Macmillan Press Ltd, London and Basingstoke, pp. 1–14 and 33–63.
    A comprehensive review, containing some useful references.
SALISBURY, E. (1961), *Weeds and Aliens*, Collins, London, p. 384.
    An easily read general introduction to the biology of weeds.

3   Weed control methods

3.1 Introduction

3.1.1 NON-CHEMICAL METHODS

In these days of near universal availability of chemicals it is possible to lose sight of the value of non-chemical means of weed control and the part which they play in minimising weed problems. Most cultivation techniques achieve some measure of weed control and some have been developed specifically for this purpose. The continuing importance of this aspect should, therefore, not be underestimated and will be referred to in later sections, as appropriate. At this stage, however, it may be as well to remind ourselves of the specific ways in which different cultivation methods actually kill weeds. These include:

(i) Burial   The main function of ploughing, apart from initial seed-bed preparation, is to bury weed seeds and other harmful fragments beneath the soil. This applies both to arable land, where seeds are the chief consideration, and to grassland, where vegetative regeneration must be minimised. Historically, ploughs have developed over a very long period from primitive, wooden 'grubber' types to more special-ised versions, introduced early this century, such as the 'digger' plough for spring work, and the 'lea' plough for maximum inversion of grass swards. Today the plough is still, overall, the main method of primary cultivation despite the introduction of alternative techniques.

As a weed-control implement, however, the plough is something of a two-edged weapon! While it is effective in burying freshly shed weed seeds, in so doing it preserves many of them by placing them at a depth in the soil where they are likely to enter a dormant phase, which may last for many years. At the same time it brings back to the surface many buried and still viable seeds from previous years, so perpetuating the problem. Hence the value of reduced cultivations and direct-drilling, which neither bury new seed nor bring back old seeds into circulation, with the result that, if further seeding can be prevented, the number of annual weeds (which depend entirely on seed for their survival) ought to be significantly reduced. There is indeed some evidence of this.

(ii) Stimulation and exhaustion   The stimulation of seeds and vegeta-tive buds into active growth is a well-tried method of producing shoots, which can then be attacked by herbicides or by further cultivation. If this is carried out repeatedly it also acts as a drain on storage reserves in rhizomes etc., reducing the weeds' capacity to recover. Various types of implement can be used for this purpose, notably rotary cultivators, which break up the rhizomes of couch-type grasses, so inducing them to form vulnerable above-ground shoots. In pre-glyphosate days this was widely adopted as a means of exhausting these grasses, prior to final deep ploughing (7.4.4).

(iii) Desiccation   This is mainly associated with the use of harrows and cultivators, both to uproot seedlings and to bring roots and rhizomes to the surface where they are left to dry out before being removed or burned. Any disturbance by pre-sowing or inter-row cultivation will have this effect, by destroying the close contact between the weeds and their moisture supply. Originally harrows and cultivators too were crude wooden implements but they have developed into a variety of sophisticated tilth-creating, tined and disc-type machines, often working in tandem and capable of one-pass seedbed preparation ('tillage-trains').

(iv) Physical destruction   This involves both the breaking-up of seedlings and plants by machinery generally and the more specific cutting and severance resulting from hand or machine hoeing. One of the greatest benefits of increased mechanisation has probably been the near-elimination of hand-hoeing. Even long after the introduction of the horse-drawn steerage hoe, hand-hoeing remained an integral part of the husbandry of many root crops, until both were replaced in post-war years, on all but the smallest units, by tractor-powered, down-the-row weeders, gappers and thinners.

3.1.2 DEVELOPMENT OF HERBICIDES

The major contribution of the twentieth century to the control of weeds has, of course, been the introduction of herbicides or chemical weedkillers. These have been in use in cereals since the early 1900s, but until the last war were few in number and mostly unpleasant, not to say dangerous, to use (Table 3.1). Research into safer and more effective materials commenced in the 1930s and resulted in the discovery, during the war years, of MCPA and 2,4-D in Britain and 2,4-D independently in the USA. These, together with propham, formed the nucleus of the present-day range of synthetic herbicides.

In the early post-war years development centred on the hormone herbicides or 'growth-regulators' in cereals – the major cash crop – but in the 1960s an increasing number of soil-applied pre- and latterly post-emergence chemicals became available for root crops, as formulation chemists gradually overcame the problems of herbicide selectivity between dicot weeds and crops. In the 1970s the emphasis was on grass herbicides, culminating in the introduction of the revolutionary 'graminicides' at the end of the decade. A full account of these

Table 3.1 *Major events in the development of herbicides before 1941*

1896	France	Copper sulphate first used for selective weed control in cereals
1896	Britain	Copper sulphate first used for selective weed control in cereals
1901–19	Europe and USA	Ferrous sulphate, sulphuric acid, sodium chlorate used as herbicides
From 1930	Britain	Substantial acreage of cereals sprayed annually with sulphuric acid
1932–3	France	Dinitro-phenols and cresols patented and used for weed control in cereals

developments is given in the *Weed Control Handbook* (Roberts, 1982).

This rapid increase in the availability of herbicides has had major effects on agriculture besides the basic matter of killing weeds more effectively. The introduction of selective herbicides gave a new lease of life to some of the traditionally labour-intensive row-crops, as shown by the rapidity with which these materials were taken up, for example for sugar beet and potatoes. Early concentration on control of broad-leaved weeds helped to produce a grass-weed explosion in the 1950s and 60s, which rumbles on to the present. The cost of chemical weed control, in cereals especially, is placing increasing strains on falling profit margins and may necessitate a return to the rotational systems which have persisted in the less intensive mixed-farming areas. At the other end of the spectrum the widening scope of herbicides in vegetable crops has influenced the use of revised growing techniques, including bed systems.

The range of herbicides available to British farmers and growers has therefore increased dramatically over the last 30 years from 14 chemicals (127 products)* in 1960, to 63 chemicals (402 products) in 1970, and 80 chemicals (437+ products) in 1982. The description, classification and utilisation of these, which have become increasingly daunting tasks for users and advisers alike, form the substance of the following sections.

*Approved under Agricultural Chemicals Approval Scheme.

3.2 Chemical methods

Before considering the principles of selectivity and the mechanisms of action of herbicides, it will be helpful to devote some attention to the classification of herbicides and the different types of current herbicide treatments.

3.2.1 CLASSIFICATION OF HERBICIDES

Herbicides may be classified according to three criteria as follows:

a) Chemical structure
b) Mode of action
c) Type of treatment

a) Chemical structure

In this method of classification, herbicides possessing a similar chemical structure or functional group are grouped together: for example, the growth regulator/auxin-type herbicides like 2, 4-D and MCPA; and the triazine herbicides which include simazine and atrazine (Fig. 3.1).
Using this procedure a number of different herbicide groups can be identified (Table 3.2).

3.2.2 MODE OF ACTION

In order to simplify a somewhat complex subject, in this chapter the term 'mode of action' will be used to describe whether a herbicide acts via the foliage or the soil, and 'mechanism of action' will refer to the way in which a herbicide interferes with the morphological, physiological and biochemical processes within the plant and so produces, ultimately, the death of the plant.

In general, herbicides are applied to the foliage and stems of plants or, alternatively, they may be applied to the soil and subsequently taken up by plant roots and shoots.

3.2.3 THE ENTRY OF HERBICIDES INTO PLANTS

a) Movement of herbicides across the cuticle

Before herbicides can inflict their phytotoxic action on weeds, they must penetrate the waxy, water-repellent membrane which covers the entire surface of the above-ground parts of all plants. This membrane, the CUTICLE, serves to protect the delicate underlying cells from the desiccating effects of dry air.

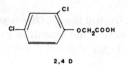

2,4 D

(2,4 dichlorophenoxyacetic acid)

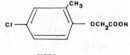

MCPA

([4-chloro-o-tolyloxy]acetic acid)

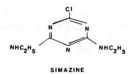

SIMAZINE

(2-chloro-4,6-bio[ethyl-amino]-1,3,5-triazine)

ATRAZINE

(2-chloro-4-ethylamino-6-isopropylamino-

1,3,5-triazine)

Fig. 3.1: *Examples of herbicide classification according to chemical structure.*

TABLE 3.2 *Groups of herbicides*

	Group	Examples
1.	INORGANIC HERBICIDES	SODIUM CHLORATE; AMMONIUM SULPHATE; BORAX
2.	HALO-ALKANOIC ACIDS	TCA; DALAPON
3.	PHENOXYALKANOIC ACIDS	2,4-D; MCPA; MECOPROP; 2,4,5-T
4.	AROMATIC ACIDS	DICAMBA; CHLORTHAL-DIMETHYL
5.	AMIDES	PROPYZAMIDE
6.	NITRILES	DICHLOBENIL; IOXYNIL
7.	ANILIDES	PROPACHLOR
8.	NITROPHENOLS	DINOSEB
9.	NITROPHENYL ETHERS	NITROFEN
10.	NITROANILINES	TRIFLURALIN
11.	CARBAMATES	CHLORPROPHAM; BARBAN; PHENMEDIPHAM; ASULAM
12.	THIOCARBAMATES	EPTC; DI-ALLATE
13.	UREAS •	MONURON; LINURON
14.	TRIAZINES	SIMAZINE; ATRAZINE
15.	PYRIDINES	PARAQUAT; DIQUAT; PICLORAM
16.	PYRIDAZINES	PYRAZONE
17.	URACILS	LENACIL; BROMACIL
18.	HETEROCYCLIC NITROGEN COMPOUNDS: UNCLASSIFIED	AMINOTRIAZOLE
19.	HETEROCYCLIC COMPOUNDS: OTHER HETERO ATOMS	ETHOFUMESATE
20	ORGANOPHOSPHORUS COMPOUNDS	GLYPHOSATE
21.	SOIL FUMIGANTS	DAZOMET, METHYL BROMIDE

The cuticle, although very water-repellent, permits slow evaporation of water. Current opinion suggests it has a sponge-like structure whose minute pores open and close as humidity rises and falls. The cuticle is also thought to be perforated by another type of pore called ECTODESMATA which may act as a site of preferential entry for herbicides (Fig. 3.2).

There appear, therefore, to be three main routes along which herbicides may diffuse across the cuticle to reach the underlying cells: an aqueous route for water-soluble herbicides, which functions efficiently only in humid conditions when the cuticle pores are full of water; a lipoidal route for fat-soluble herbicides; an ectodesmatal route for all herbicides.

Stomata are often considered important sites of entry for herbicides, but there are few, if any, stomata on the upper surface of a leaf, the surface which receives the bulk of a spray. Even if a herbicide enters a leaf via stomata, it must still pass through the thin cuticle lining the substomatal cavity before entering the leaf cells. It is not surprising, therefore, that under field conditions relatively little herbicide enters the plant via this route.

b) Movement of herbicides within the plant to the sites of action

Following penetration of the leaf cuticle, herbicides must gain access to the long-distance transport system of the phloem if they are to destroy tissues remote from the area of contact by the spray. This requires their transport into the epidermal cells followed by cell to cell diffusion to the tissues surrounding the phloem, and finally loading into the conducting vessels themselves. Since energy from respiration is needed at some point in this pathway, herbicides which act by inhibiting respiration should not be applied at high concentrations otherwise they will not reach the conducting vessels in sufficient quantity to ensure effective distribution in the long-distance transport system.

Once inside the phloem vessels, the rate and pattern of herbicide distribution is similar to that of sugars. The rate of movement is greatest in young, vigorously growing plants, which explains why weeds should be sprayed during phases of rapid growth. Herbicides entering young leaves remain there and are not exported downwards to older leaves and roots. Unfortunately, the bulk of a spray usually falls on these young, non-exporting leaves. Older leaves can and do export herbicides to underground organs but they often receive only a small proportion of the total volume of a spray.

c) Entry of soil-applied herbicides into plants and their transport to the shoot

Soil-applied herbicides move to the root surface mainly by mass flow of

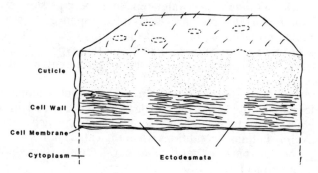

Fig. 3.2: *Section of part of a leaf cell showing the spongy waxy cuticle and ectodesmata.*

the soil solution which takes place in response to transpiration. Absorption by the cells of the roots, and underground parts of shoots, is followed by migration to the xylem vessels and upward transport to the shoot. Some herbicides carried to the leaves in the xylem sap may also be exported via the phloem, thus ensuring an effective distribution throughout the plant.

Broadly speaking, herbicides may be divided into two classes with respect to mode of action, i.e. LEAF ACTING and SOIL ACTING. Leaf-acting herbicides can be further categorized into CONTACT herbicides which exert their effects only where they touch the plant, and TRANSLOCATED herbicides which are taken into the plant and transported, in either the phloem or the xylem, to their sites of action, which are usually regions of active growth, reproduction or storage (Fig. 3.3).

Both types of leaf-acting herbicides have associated advantages and disadvantages. Contact herbicides, e.g. paraquat, are preferable since they kill weeds very quickly and, as there is no persistence, a crop can be planted soon after treatment. The disadvantages are that good cover of the weed by herbicide spray is essential; normally only annual weeds are controlled – deep-rooted perennials are only checked and there is no long-term control.

The benefits of leaf-acting translocated herbicides like 2,4-D and glyphosate are that since the chemical accumulates at the site of action good spray cover is not critical; also that, because the herbicide is transported within the plant, even well-established perennial weeds are controlled. However, care is required when applying these chemicals in order to avoid spray drift and subsequent damage to adjacent susceptible crops. Moreover, for maximum control, timing is critical and there is again no long-term control of weeds which develop from seeds or propagules after spraying.

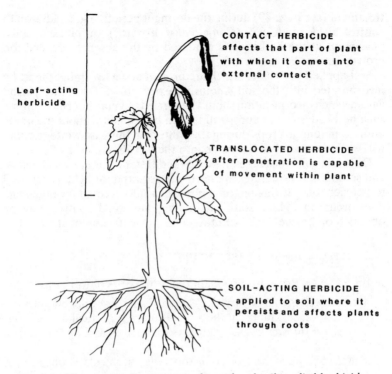

Fig. 3.3: *Illustration of contact, translocated and soil-applied herbicides (courtesy D. W. Robinson).*

Soil-acting or soil-applied herbicides are taken up by plant roots and then transported to their sites of action within the plant (Fig. 3.3).

The advantages of this type of herbicide are that they have a long-lasting effect, remaining active in the soil until broken down or washed out of the region of the soil where they exert their effects; and also that they are relatively easy to apply, either as granules which bounce off plants on to the soil surface, or as sprays. Granular herbicides also facilitate 'placement', in that the herbicide can be applied to the exact region where it is required.

These herbicides are, however, very dependent on soil water. If the soil is too dry they are not effective because they are not dissolved in soil water and cannot therefore be taken up by plant roots. Alternatively, if conditions are too wet there is a danger of crop damage, as the soil water may carry the herbicide to the crop roots. Also, soil-applied herbicides have a broad spectrum of activity and under certain circumstances crop plants are fully as susceptible as weeds to these chemicals. Moreover, where a soil-applied herbicide is used as a spot treatment (see page 49) during the dormant period, e.g. oxadiazon to control field bindweed (*Convolvulus arvensis*) in blackcurrants, placement of the herbicide is hindered by the absence of weed top growth during the winter.

Soil-applied herbicides which are broken down by sunlight must be incorporated into the soil, requiring extra effort. Also, the period within which incorporation should be carried out may be critical. Some such herbicides (e.g. napropamide) can be applied without incorporation to perennial crops during the winter. In this case winter rainfall will wash the chemical herbicide into the top 50mm layer of soil.

Different soil types take up (ADSORB) different amounts of chemical and so different dose rates are normally required for light as opposed to heavier soils. If this factor is not taken into account crop damage may occur on lighter soils and inadequate weed control may be obtained on heavier soils with the same concentration of chemical.

3.2.4 TYPES OF HERBICIDE TREATMENTS

In broad terms there are two main types of weed control as follows:

i) *total weed control* In this case herbicides are applied with the intention of killing all the vegetation in a particular situation, e.g. on railway tracks, pathways and driveways, and industrial areas.

ii) *selective weed control* With this type, herbicide application is made with the objective of controlling certain plants (weeds) while, at the same time, causing no harmful effects to others (crops or ornamentals). Situations where selective weed control is used include grassland, vegetables, tree and shrub beds.

Selective use of herbicides in crop situations

A crop that is raised from seed can exist in three stages – not drilled; drilled but not emerged; drilled and emerged. A crop established from plants (e.g. transplanted cabbage, strawberries, apples) exists only in the emerged stage. The third type of herbicide classification therefore relates to the stage of development of the crop at which the herbicide is applied. Three possibilities exist:

a) *Pre-sowing herbicide treatments* in which the herbicide is applied before the crop is drilled.

b) **Pre-emergence herbicide treatments* are carried out where the crop has been sown but has not emerged through the soil.

c) **Post-emergence herbicide treatments* are employed when the crop has emerged above ground. In each case weeds may, or may not be present above ground.

(*The terms pre- and post-emergence refer to the crop, *not* the weed.)

In each of these types of herbicide treatment, soil-applied and foliage-applied herbicides may be used either alone or in combination. The type of herbicide or herbicides employed depends on a) whether a crop is present, b) whether the weeds have emerged, and c) the types of weeds, if any, which are present.

The types of herbicide treatment discussed so far normally involve spraying the chemical over the entire crop or other area to be treated. There are occasions, however, when a herbicide may be too expensive to warrant such a treatment, or weeds may be present only in small numbers or in isolated patches. In these circumstances treatment of the whole crop or area would be both undesirable and unnecessary, and other types of herbicide treatments might be considered. DIRECTED SPRAYING, which may be used in tall, widely-spaced crops, involves treating the soil or weeds, if present, with herbicide spray while avoiding contact with the crop plants. Alternatively, in BAND APPLIC-ATION, which normally involves the use of pre-emergence soil-applied herbicides, the chemical is sprayed only over the crop row, and weeds between the rows are destroyed either by manual or mechanical means. Where only a few weeds are present or where problem weeds are growing within a crop – e.g. thistles in strawberries – these may be treated individually. This is termed SPOT TREATMENT. The different herbicide treatments are shown in Fig. 3.4.

3.2.5 HERBICIDE SELECTIVITY

Herbicide SELECTIVITY is taken to mean the situation where a chemical reaches and disrupts a vital function in one plant, or group of plants, and not in another.

In practice there are a number of ways in which selectivity may be achieved and, in general, several mechanisms may be involved.

For simplicity, it is convenient to divide the various selectivity mechanisms into non-physiological and physiological categories.

Non-physiological mechanisms

There are several possibilities for using a herbicide in crops which are fundamentally sensitive, including the use of suitable methods of application. For example, the herbicide may be applied so that the spray does not touch the crop plant, or comes into contact only with its insensitive base. This is the case when paraquat is used as a directed spray on low-growing annual weeds at the base of taller-growing shrubs, or when it is applied to weeds to clean up beds of narcissi after the bulb foliage has *completely* died down.

In addition, the behaviour of herbicides in the soil can be utilised. Generally, in comparison with crop plants, weeds germinate and develop their roots relatively close to the soil surface. Certain soil-

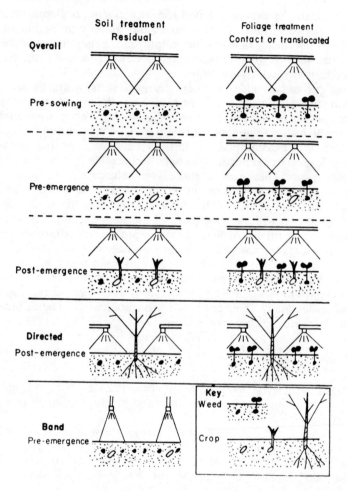

Fig. 3.4: *Definition of herbicide treatments (after Fryer and Makepeace, 1977).*

acting herbicides can therefore be used selectively in crops because they are retained close to the surface. Thus they kill weed seedlings with shallow root systems, but they do not normally come into contact with the deeper roots of crop plants. This is called POSITIONAL SELECTIVITY or DEPTH PROTECTION. While this may occasionally provide a useful means of selectivity it is not reliable in all circumstances, since not all crop plants are deep-rooted nor all weeds shallow-rooted. Moreover, in very wet conditions and on light soils there may be a danger of movement of herbicide into the region where the crop roots are to be found. Tolerance of herbicides in crop and ornamental plants is not due entirely to depth protection, as seeds of some ornamentals including *Cotoneaster wardii*, *Erica arborea* and *Berberis darwinii* have been found germinating freely in simazine-treated soil.

Depth protection may also be found with herbicides which penetrate the shoot. The control of wild oat (*Avena fatua*) in wheat, using tri-allate, is based on the knowledge that the growing point of the wild oat quickly enters the herbicide region if the chemical is only shallowly incorporated, while the growing point of the wheat only comes into contact with the herbicide layer later, when it is shielded by other tissues.

Morphological differences between plants may result in selective action by herbicides, by affecting spray interception and retention and absorption of the herbicide through the cuticle. In cereals, for instance, the narrow erect leaves provide less of a target than the broad flat leaves of many weed plants and, consequently, intercept less spray. Similarly, dilute sulphuric acid, once widely used as a herbicide in onions, caused little damage to this crop because the upright leaves and waxy cuticle shed most of the spray, whereas the spray was retained by the hairy leaves of broad-leaved weeds. It is important to appreciate, however, that not all hairy leaves shed herbicide spray. the leaves of tomato, for instance, although hairy, readily retain spray.

The important factor in determining whether spray is retained is not the hairiness or smoothness of the leaf, but the shape and configuration of the microscopic wax particles on the leaf surface. For instance, the mealy surface of the leaves of fat hen (*Chenopodium album*), which is due to wax globules, is not nearly so water-repellent as the type of wax found on pea leaves, so this weed can be controlled in peas by a contact herbicide like dinoseb.

In addition, spray retention is dependent on the age of the plant. The cuticle in young plants is not complete, which generally means that seedlings retain spray more readily than mature plants. However, leaves of older plants can again become retentive if the cuticular wax is damaged, since they are unable to renew the wax once it has been destroyed.

Furthermore, the production of cuticular wax tends to be reduced under warm humid conditions which could lead to crops retaining a greater amount of herbicide spray and, consequently, being more susceptible to damage under these conditions.

The addition of wetting agents to herbicide sprays can also affect spray retention and selectivity. Too much wetting agent or the wrong wetting agent can influence the spray in two main ways. It may reduce the surface tension to such an extent that excessive run-off occurs, resulting in unsatisfactory weed control. Alternatively, it may facilitate retention and cuticular penetration so that crops which would normally reflect the spray may be damaged.

Retention of herbicide spray is also affected by droplet size and the volume of spray applied. Broadly speaking, low-volume high-pressure sprays produce a high proportion of small droplets (a fine spray) which tend to be retained by leaves, while high-volume low-pressure sprays produce a high proportion of large droplets (a coarse spray) which tend to bounce off leaves (4.2).

Furthermore, the selectivity of a herbicide like sulphuric acid, on onions and related crops, is enhanced by the position of the MERISTEM or growing point. Broad-leaved (dicotyledonous) weeds usually have an exposed growing point which is easily wetted and destroyed by a contact herbicide. In comparison, the growing point in onion is basal and is therefore protected from the spray by leaf sheaths.

Physiological selectivity

Selectivity due to physical and morphological factors often depends on conditions outside the control of the herbicide user. Heavy rain, for instance, may wash a soil-acting herbicide more deeply into the soil than usual, or a prolonged period of humid weather may result in crop plants having a thinner than normal cuticle, which would make them more easily damaged by certain contact herbicides.

On the other hand, physiological selectivity is a considerably more constant and certain mechanism, as it makes use of basic differences between weeds and crop plants. These include differences in absorption and translocation to sites of action and differences in degradation processes.

A number of factors may result in differential absorption of herbicides by crop plants and weeds, including differential exposure of the roots and green parts as a consequence of different growth patterns, and differences in the anatomical and chemical structure of the surface tissues of the plants.

Figure 3.5 illustrates some of the barriers, including absorption, which may prevent a herbicide from reaching its site of action and subsequently exerting its effects.

Differences between weeds and cultivated plants in the rate of translocation of herbicides appear to be a common selectivity mechanism. For example, part of the tolerance of grasses to 2,4-D is due to slow translocation of the chemical. A herbicide may also be non-toxic because it is absorbed by plant tissues and is therefore rendered unavailable for further translocation. Alternatively, a herbicide may be broken down (detoxified) by plant enzymes as it moves through the plant.

There is evidence that the selective action of herbicides such as 2,4-D, MCPA, simazine and atrazine is due to rapid and effective degradation processes in some cultivated plants. In maize, for instance, simazine and atrazine are converted into a non-phytotoxic

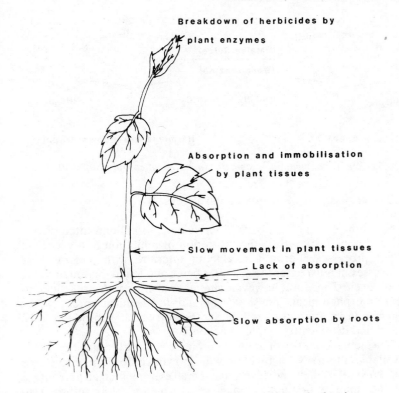

Breakdown of herbicides by plant enzymes

Absorption and immobilisation by plant tissues

Slow movement in plant tissues

Lack of absorption

Slow absorption by roots

Fig. 3.5: *Some of the barriers which may enable a herbicide to act selectively (courtesy D. W. Robinson).*

breakdown product by the removal of the chlorine atom and its replacement by a hydroxy (-OH) group (Fig. 3.6).

Studies have shown that the breakdown of simazine and atrazine by maize is due to the presence of a substance known as BENZOXAZINE and, while this is not the full story, plants sensitive to chemicals like simazine generally have a lower benzoxazine content compared with maize.

In contrast, some plants are apparently resistant to herbicides as a result of their inability to break down the chemical. The best example of this type of selectivity is the use of MCPB for weed control in cereals undersown in clover. Most plants, with the exception of legumes including clover, convert MCPB, which is harmless, into MCPA, which is phytotoxic, and so destroy themselves. Clover, however, is unable to carry out this conversion, and so remains unharmed.

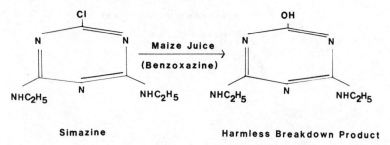

Fig. 3.6: *Detoxification of simazine in maize (after Dubach, 1970).*

3.2.6 MECHANISMS OF ACTION OF HERBICIDES

The ultimate aim of applying a herbicide is to kill unwanted plants. This may be achieved by the disruption or inhibition of a variety of plant processes. For a herbicide to interfere with one of these processes, it must enter the cell after being applied to the soil or to the plant itself. There are, however, a number of environmental, anatomical, morphological, physiological and biochemical forces which operate in an attempt to counteract the entry of a herbicide into the plant and its reaching its site of action.

The actual quantity of applied herbicide which eventually becomes available internally for reaction with growth-controlling processes is unknown. It is likely, however, that it is an extremely small percentage of the amount originally applied. Nevertheless, it is sufficient to impede and derange cellular processes to such an extent that the plant is eventually killed.

It is difficult to provide a simple overall picture of the mechanisms of action of herbicides. While some herbicides may have only one primary site of action, it is likely that others interfere with a range of plant processes. This makes investigation of mechanisms of action difficult as it is no easy matter to distinguish between direct and indirect effects.

Before considering the biochemical aspects of herbicide action we shall look at some of the morphological and physiological effects on plants which result from herbicide treatment.

Most herbicides may be grouped on the basis of whether they affect germination, including the initial growth of the seedling, or whether they act at a later stage in the development of the plant due to their absorption from the soil or through treated leaves and stems.

1. Herbicides which disrupt germination and initial growth of the seedling (germination inhibitors)

Germination inhibitors are usually compounds in the classes carbamates (e.g. chlorpropham), thiocarbamates (e.g. EPTC), anilides (e.g. propachlor) and nitroanilines (e.g. trifluralin).

Herbicides which inhibit germination may also act on the later stages of plant growth, although their effect is greatly reduced and of no practical value. Furthermore, they may initially stimulate germination, only to act more vigorously upon the young seedling.

2. Herbicides which act after the seedling stage

Chemicals in this category may be grouped according to the morphological effects which they produce, and in order to simplify the situation, a distinction can be made between the following:

a) *Burning*: This consists of rapid corrosion at the primary point of contact of the herbicide with the plant. This may spread, but it is not preceded by bleaching of the green parts (i.e. CHLOROSIS). Burning is produced by herbicides like paraquat, ioxynil and bromoxynil.

b) *Chlorosis*: In this case, the green parts of the plant are progressively bleached, and become white and die relatively quickly. Aminotriazole acts in this manner, possibly by blocking the formation of the chloroplasts.

c) *Gradual chlorosis which progresses to necrosis (browning or blackening of the tissues) leading eventually to the death of the plant*: This is the mode of action of herbicides which inhibit photosynthesis.

On the whole, plants and animals have roughly the same kind of biochemical metabolism, but one important difference is that green plants, with the aid of the sun's energy, are able to convert carbon dioxide (CO_2) from the air into chemical energy which is stored in the form of carbohydrate compounds.

Today we have a very good understanding of the individual biochemical steps of photosynthesis. Chlorophyll molecules which are located in the chloroplasts convert light energy into unstable energy-rich compounds like ATP (ADENOSINE TRIPHOSPHATE) and NADPH (NICOTINAMIDE ADENINEDINUCLEOTIDE PHOSPHATE REDUCED – Reduced in this case means H has been added) which then bring about the first steps in the incorporation of carbon dioxide into organic compounds.

Detailed knowledge of the individual steps in photosynthesis has enabled us to identify the sites of action of many of the herbicides which inhibit this system. One such process is the Hill reaction, or the liberation of oxygen from water (Fig. 3.7), which is inhibited by, among others, the triazine, urea and uracil-type herbicides. Another well-known site of action is at 'light reaction 1' where the formation of NADPH is inhibited by paraquat and diquat (Fig. 3.7).

d) *Inhibitors of respiration*: Besides synthesising energy-rich compounds like sugars and starch, plants also liberate and utilise stored energy to perform vital functions such as growth and development.

In respiration, which takes place within plant cells in organelles

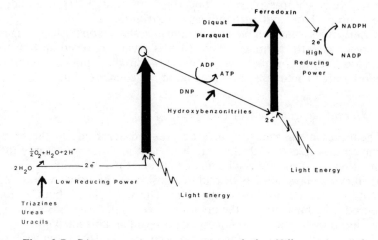

Fig. 3.7: *Diagrammatic representation of the Hill reaction (after Dubach, 1970).*

called the MITOCHONDRIA, carbohydrates (assimilates) are burned (oxidised) in the presence of oxygen, and ATP is produced which, in turn acts as an energy source or 'motor' for the numerous vital processes in the plant. Typical respiration inhibitors like dinoseb, ioxynil and bromoxynil inhibit the formation of ATP, as such, so that the plant eventually respires itself to death.

e) *Inhibitors of cell division*: From a biochemical point of view, the products described previously as germination inhibitors should strictly be termed cell division inhibitors. Interference with cell division or mitosis eventually leads to the cessation of growth of shoots and roots. The enzyme α-amylase is essential during germination for the mobilisation of carbohydrate reserves required for the growth and development of the embryo, and it has been found that certain herbicides, for example propachlor, block the production of this enzyme.

f) *Intervention in hormone systems*: The most important herbicides in this grouping are the auxin-like chemicals 2,4-D and MCPA. At low concentrations these herbicides mimic the action of the natural auxin IAA but, whereas IAA levels are probably controlled by the enzyme IAA-oxidase which breaks down IAA, there is little or no regulation of the level of synthetic compounds, and this leads to their herbicidal activity.

Present evidence suggests that this type of herbicide may interfere with nucleic acid metabolism. Moreover, their action may involve an effect on cell membranes or cell walls or, indeed, there may be a parallel action at both of these sites.

g) *Inhibition of the biosynthetic processes of the cell*: The effects of herbicides on photosynthesis and respiration are, undoubtedly, important mechanisms of action, but there is increasing evidence that certain herbicides exert their effects at a more basic level, for example on protein synthesis. Glyphosate is thought to interfere with the formation of certain amino acids (the building blocks of proteins) and any disruption of their formation would inevitably effect protein synthesis. Although the exact mechanism of action of glyphosate has not, as yet, been fully ascertained, it is possible that this chemical influences the activity of the ribosomes, which are essential to protein formation.

Some herbicides, including EPTC and di-allate, are believed to interfere with the formation of fatty acids which are the major components of plant lipids (fats). Lipids are important constituents of cell membranes, and any interference with their synthesis would affect the function of the membrane, which is of crucial importance in controlling the flow of materials into and out of the cell. Furthermore, the waxes in the cuticle are composed of fatty acids, and any decrease in the formation of the cuticle would render the plant less able to withstand attack by pests and diseases, and would lead to increased water loss. Under certain circumstances, dalapon is known to produce a reduction in cuticular wax.

It is evident, therefore, that herbicides may act on a wide variety of plant processes, but despite the volume of research in this field, there are still large gaps in our knowledge concerning the detailed mechanisms of action of these materials.

3.3 Loss, breakdown and persistence of herbicides

Following application to either the foliage of plants or to the soil, herbicides may be lost in several ways (Fig. 3.8).

3.3.1 LOSS FROM THE PLANT

Herbicides may intially disappear as a result of volatilization or photodecomposition, and this is particularly true of those materials which are retained on or close to the surface of leaves and stems. Additionally, within the plant, herbicides may be degraded by a variety of biochemical processes to form non-toxic compounds, and so lose their herbicidal properties. Moreover, herbicides may be lost from plants quite simply when leaves become detached either as a result of ageing or of herbicidal injury. A number of foliage-applied herbicides may, in fact, be lost via the roots of treated plants, but the significance of loss by this method is not known at present.

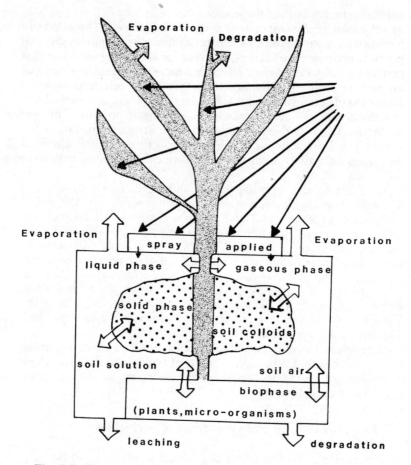

Fig. 3.8: *Processes by which herbicides are lost from plants and soil (after Dubach, 1970).*

3.3.2 HERBICIDES IN THE SOIL

Before considering the ways by which herbicides are removed from the soil, it is appropriate to give some attention to the interactions which take place between these chemicals and the different systems which exist within the soil environment.

The behaviour of herbicides in the soil is of considerable importance, not only in the case of soil-applied herbicides which are applied directly to the soil, but also with foliage-applied herbicides since a proportion of the active ingredient inevitably reaches the soil, either by bouncing or running off treated plants, or by missing the target plants completely.

The soil exerts a strong influence on the activity of soil-applied herbicides as they encounter a dynamic situation involving interactions between the herbicide and the individual components of the soil. These include clay, humus, the soil solution, the soil air, plant material and soil organisms.

These interactions may be divided into three main groups as follows:

First. The herbicide undergoes a distribution process whereby it may be dissolved in the soil solution, *adsorbed* on to the solid COLLOIDAL* soil bodies such as clay and humus, *evaporated* in part into the soil air, or taken up (*absorbed*) by plants and micro-organisms. [*A COLLOID is a system in which tiny particles are dispersed in some medium. The SOIL COLLOIDAL SYSTEM consists of solids (clay and humus) dispersed in a liquid (the soil solution).]

Second. Within the soil the herbicide may be transformed or degraded by various chemical and biochemical processes.

Third. By virtue of its biological activity, the herbicide may exert an influence on the soil itself by affecting soil fertility for example.

The Soil System

The soil is a dynamic, complex system which contains gaseous, liquid, solid and living components or PHASES. The solid phase is, to a large extent, present in a very finely distributed (COLLOIDAL) form which means that it provides an extremely large surface area.

For clarification, if we start with a 1cm cube of solid material the total exposed surface area is $6cm^2$ or $600mm^2$. If we subdivide the cube into smaller cubes of 1mm (about the size of coarse sand), this produces one thousand 1mm cubes with a total surface area of $6,000mm^2$. To obtain an idea of the surface area provided by colloidal clay, we would have to divide the original 1cm, cube into cubes with sides of 0.001mm. This would give us 10^{12} such cubes with a total surface area of 60 million mm^2. Therefore, for a given volume, a vast increase in total surface area can be obtained by continuous sub-division.

In soils, only the clay particles are colloidal in size, and even some of these may be too large for this category. Table 3.3 illustrates how the surface area per gram of solid mass increases as the diameter of the solid particle decreases.

In addition to inorganic colloids (i.e. clay) the soil also contains organic colloids or HUMUS derived by the breakdown of organic materials in the soil by micro-organisms.

The molecules which make up the soil colloidal system are not electrically neutral but are, in the main, negatively charged. This is extremely important in relation to the behaviour of herbicides in the

TABLE 3.3   *Increase in surface area with decrease in particle diameter*

Fraction	Particle diameter (mm)	Approximate surface area (cm^2/g)
Coarse gravel	200–20	–
Fine gravel	20–2.0	–
Coarse sand	2.0–0.2	21
Fine sand	0.2–0.02	210
Silt	0.02–0.002	2,100
Clay	0.002	23,000

soil. Once they have entered the soil system, herbicides are exposed to the electrostatic attractive forces (i.e. the pull of the negative charges) of the surfaces of the soil colloids, and are ADSORBED on to these surfaces, which makes it more difficult for them to be taken up by plants. Adsorption may be thought of as similar to the attraction of iron filings to a magnet, the soil colloids being the magnets, and the herbicides the iron filings.

Although good correlations have been obtained in laboratory experiments between herbicidal activity and soil characteristics such as colloid content, under field conditions climatic factors apparently exert a greater influence on the behaviour of herbicides than the soil factors themselves.

The process of adsorption is reversible, which means that herbicide molecules do not, as a consequence of adsorption, vanish into the soil colloid and accumulate there. If the herbicide concentration in the soil solution decreases either as a result of uptake by plants or by degradation, herbicide molecules detach themselves from the soil colloids until the concentration of herbicide in the soil solution is restored to its original state.

Having considered briefly the theoretical aspects of adsorption we may now examine the more practical aspects of the behaviour of herbicides in the soil.

Leaching

Leaching may be described as the movement of a chemical as influenced by the flow of soil water and, for all practical purposes, may be considered as movement downwards in the soil, although sideways and upwards movement does occur.

Leaching is extremely important with regard to herbicide activity, since, in order to achieve maximum weed control, soil-applied herbicides must normally be washed into the soil so that they reach the region where weed roots and germinating weed seeds are to be found.

Certain herbicides such as trifluralin are not readily washed into the soil by rain, and thus they may fail to come into contact with the germinating weed seeds. They may therefore be exposed to high solar radiation, and may be broken down by sunlight (PHOTOCHEMICALLY DECOMPOSED) or evaporated. Such herbicides have therefore to be incorporated into the soil.

Too strong a tendency towards leaching, as is found with picloram, may lead to the emergence of shallow-rooted weeds, contamination of drainage water and injury to the deeper-rooted crop plants. Such chemicals are useful, however, for the control of deeply-rooted weeds. From a practical point of view, the greater the leaching properties of a herbicide, the higher are the demands made upon its physiological selectivity in relation to the crop.

Leaching is increased by a number of factors including the adsorption capacity of the soil. The smaller the adsorption capacity, i.e. the fewer the colloidal soil components present, and the more the negative charge of these particles has been neutralised by a high soil pH, the greater will be the tendency towards leaching. Moreover, leaching tends to increase with increasing frequency and level of rainfall.

Evaporation

The distribution or movement of a herbicide through the gaseous soil phase (soil air) and into the atmosphere can lead to serious loss of chemical and damage to crops.

Volatility varies with different herbicides, and some products must be worked into the soil soon after application to prevent them escaping into the atmosphere as vapour before they exert their desired effect.

Evaporation, like leaching, increases as a) the adsorption capacity of the soil decreases, b) rainfall increases, and c) temperature rises. Evaporation is particularly likely to occur when there is continuous light rain accompanied by high temperature, and under these conditions herbicides which are prone to evaporation generally vaporise fairly quickly. On the other hand, intensive irrigation and heavy rainfall tend to produce the same effect as mechanically working the herbicide into the soil.

3.3.3 ABSORPTION OR UPTAKE BY PLANTS

The absorption of a herbicide into the biophase (i.e. uptake by plants and micro-organisms) is dependent upon the absorption capabilities of the organisms, the adsorption capacity and moisture content of the soil, and the currently prevailing climatic conditions. In situations in which the adsorption capacity of the soil is very high, for example in soils rich in organic matter, there may be virtually no herbicide activity.

In practice the rate of application of a herbicide may have to be altered to take account of soil type and, generally, a lower dose is required on light sandy or gravelly soils than on soils rich in organic matter. This is one reason why farmers and growers should be aware of the soil types on their holdings, since marked variation can occur, not only from field to field, but even within the same field.

3.3.4 TRANSFORMATION AND DEGRADATION OF HERBICIDES IN THE SOIL

In theory, herbicide molecules are subject to degradation in all the soil phases, i.e. biochemical processes in the biophase, and a number of purely chemical processes including oxidation, reduction and hydrolysis in the solid, liquid and gaseous soil phases.

In order to achieve optimum weed control, herbicides should possess, within certain limits, as long an effective life in the soil as is possible. Herbicides which are degraded too quickly give unsatisfactory weed control while those which are too persistent can cause damage to susceptible following crops.

The requirements as to the effective life of herbicides differ according to the situation for which they are required. In vegetables, for example, chemicals which persist for only a short time are desirable, whereas in non-crop situations like industrial installations herbicides with a much longer persistence are generally preferred.

The use of herbicides which persist in the soil began mainly in fruit and other woody perennial crops, but their use is now widespread in agriculture and horticulture. The high cost and increasing shortage of labour, coupled with a very high degree of weed control, make soil-applied herbicides an attractive proposition. Furthermore, the use of soil-applied herbicides overcomes the need for cultivation in many crops, and this in turn leads to a reduction of damage to the surface roots of crops, and to increased production. There are, however, a number of potential problems relating to the use of persistent herbicides, such as the possibility of a build-up of toxic residues in the soil, and the likelihood and results of overdosing. The attitude towards the persistence of pesticide chemicals varies. In the past, residual or persistent insecticides were considered beneficial as they could control harmful soil insects over the whole season. On the other hand, residual herbicides which have too long a persistence can produce adverse effects on susceptible following crops within a rotation.

There is a wide range in persistence of herbicides in the soil, from about four weeks in the case of 2,4-D to eighteen months with picloram (Figure 3.9). The persistence of an individual chemical varies with soil and climatic conditions, and the results in Fig. 3.9 represent the time taken for a 75–100% loss of chemical to occur at normal application rates, and under normal agricultural conditions. It is important, therefore, when choosing a herbicide to select one which

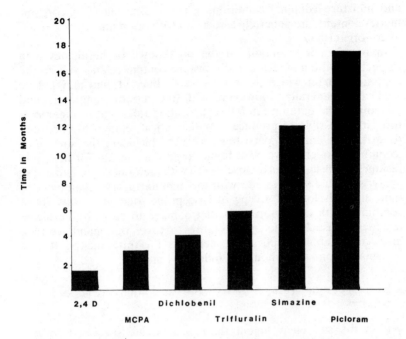

Fig. 3.9: *Persistence in soils of selected herbicides (adapted from Kearney, Nash and Isensee, 1969).*

will be efficient in controlling weeds, but whose persistence will not result in damage to following crops.

Although herbicides are lost from the soil in several ways, there is little doubt that the main factor influencing their persistence is degradation in the soil. Furthermore, degradation by soil micro-organisms is probably one of the most important routes whereby non-volatile herbicides are dispelled from the soil. Virtually all herbicides can be broken down by soil micro-organisms sooner or later. Several factors influence the rate of degradation in the soil. These include the ease with which the herbicide molecule itself can be broken down by the enzymes of soil micro-organisms, soil type, temperature and moisture content. While the interrelationships between different soil properties (e.g. pH and organic matter content) present a problem in identifying the effect each has on the rate of herbicide degradation, there is evidence to suggest that the organic matter content of soil is an important factor.

The effects of soil temperature and soil moisture content on the rate of breakdown of herbicides are, however, more readily determined. As a rule, the rate of breakdown increases with increasing temperature and moisture content, conditions which, along with high organic matter content, are generally associated with optimum activity of soil micro-organisms.

In practice, information on the breakdown of herbicides with respect to soil temperature and moisture content relates only to the conditions under which the results were obtained, and may not be applicable generally. However, with the aid of computers and meteorological data on rainfall, evaporation rates and soil temperature, it is possible to produce a mathematical relationship between these factors and the persistence of a herbicide in the soil. Such 'computer models' have been found to produce results for herbicide residues which agree fairly closely both with the values obtained by the analysis of field samples and with the manufacturers' recommended time intervals for the sowing of susceptible following crops. This, combined with the observation that damage to crops by herbicide residues is usually related to failure to carry out manufacturers' instructions about application, helps to illustrate the point that information on labels should be followed rigidly.

3.3.5 EFFECTS OF HERBICIDES ON SOIL MICRO-ORGANISMS AND SOIL STRUCTURE

Within the soil environment there exist many species of micro-organisms which, by virtue of their enzyme systems, break down organic plant and animal material into a simpler form which can then be utilised both by themselves and by developing plants. This conversion of material from the organic to the inorganic form, of nitrogen for example, is termed 'mineralisation'. In addition, certain micro-organisms like the bacterium *Rhizobium*, located within the nodules of leguminous plants, have the capacity of 'fix' atmospheric nitrogen. That is, they convert atmospheric nitrogen into a form in which it can be utilised by the plants.

Although it is not always apparent, micro-organisms provide an essential contribution to man's well-being. With regard to soil fertility they have an integral part to play in the recycling of nutrients. Additionally, they contribute to the maintenance of soil structure through the weathering of inorganic soil components, and by cementing and binding together soil particles both by means of polysaccharides which they produce, and by the hyphae of soil-borne fungi.

It is of the utmost importance, therefore, that any toxic effects soil micro-organisms sustain from the application of herbicides, particularly in the long term, are identified and eliminated.

In order to determine whether soil micro-organisms are, in fact, adversely affected by agrochemicals, including herbicides, studies have been made of various microbiological processes within the soil, including mineralisation, respiration, breakdown of organic matter, soil enzyme systems, and populations of micro-organisms in the soil.

To date the majority opinion favours the view that there are no significant harmful effects on soil micro-organisms or on soil fertility from the long-term use of herbicides, even where the doses applied are greatly in excess of those used in practice. Moreover, laboratory experiments in which *Rhizobium* was treated with the herbicides atrazine and pyrazone suggest that there is no substantial inhibition of either nitrogen fixation or respiration, and that any reduction in the numbers of bacteria produced would be unimportant in the field. In addition, if it is accepted that crop yield and quality provide an accurate reflection of soil fertility, then the available information, although somewhat limited, points to the view that, in general, there is no reduction in either of these parameters as a result of the repeated annual application of herbicides in long-term field trials.

In the past, mechanical cultivation was normally employed in most crops, for the preparation of a seed-bed, the maintenance of a loose soil structure, and the control of weeds. However, as efficient soil-acting herbicides became available, an increasing acreage of land became subject to management by the use of these chemicals with little or no cultivation. While minimum or zero-cultivation was adopted to a much greater extent in horticulture, particularly in the management of bush and cane fruit plantations, an ever increasing area of agricultural crops is now being direct-drilled, and the use of herbicides has been shown to be of benefit in most crops.

Although zero-cultivation and the use of soil-applied herbicides have only been practised for a relatively short period of time, data obtained over the past fifteen to twenty years suggest that the long-term use of herbicides may not have any serious effect on soil stability, structure and fertility. Although soils treated with herbicides over a long period have different properties from those under cultivation or under grass, they do not seem, at present, to be substantially worse for the plant. Indeed, trees and shrubs are found to grow as well, if not better, in herbicide-treated soil as in cultivated soil, because damage to surface roots is avoided when cultivation is eliminated.

While present evidence suggests that, on balance, the long-term use of soil-applied herbicides does not adversely affect the soil environment, new products, formulations, herbicide mixtures and programmes are continually being introduced. This being the case, it is essential that studies in this area of herbicide research should continue, ih order to elucidate and avoid any hazardous side effects which might occur in the future.

Further Reading

ANDERSON, W. P. (1977), *Weed Science: Principles*, West Publishing

Company, St Paul, New York, Boston, Los Angeles and San Francsico.

> A thorough but readable text, although most examples quoted refer to the American situation.

DODGE, A. D. (1977), 'The mode of action of well-known herbicides', in *Herbicides and Fungicides – Factors affecting their Activity*, McFarlane, N. R. (ed.), 1977, Alden Press, Oxford, London and Northampton.

DUBACH, P. (1970), *Dynamics of Herbicides in the Soil* and *The Mode of Action and Principles of Selectivity of Herbicides*, Ciba-Geigy Ltd, Basle.

> Two clearly-written booklets on topics of basic importance in herbicide application.

FRYER, J. D. (1981), 'Herbicides: Do they affect soil fertility?', *Span*, 24, (1), 5–10.

> A useful, easily-read article.

GREAVES, M. P. (1979), 'Long-term effects of herbicides on soil micro-organisms', *Ann. appl. Biol.*, 91, 129–32.

ROBERTS, H. A. (ed.) (1982), 'The Evolution of Weed Control', *Weed Control Handbook*, 7th edn., Blackwell Scientific Publications, Oxford, pp. 37–63.

> Valuable account of developments in weed control since early times, including glossary of herbicide introductions. Pp. 68–157 contain a first-class comprehensive review of herbicides in plants and in the soil, and a detailed account of the properties of the different groups of herbicides.

STEPHENS, R. J. (1982), *Theory and Practice of Weed Control*, The Macmillan Press Ltd, London and Basingstoke, pp. 75–115 and 130–43.

> Detailed account of herbicides, their modes of action and selectivity, plus information on the effects of herbicides on soil structure and fertility, on soil organisms and on the weed flora.

4 Formulation and Application of Herbicides

4.1 Formulation

4.1.1 AQUEOUS FORMULATIONS

A fundamental problem with herbicides, as with other crop protection chemicals, is how to spread a very limited quantity of active ingredient (a.i.) over a large area of soil or foliage, with maximum damage to the weeds and minimum to the crop, if also present. These aims are achieved by bulking up the active ingredient with an inert carrier substance, usually water, together with a variety of additives designed to improve formulation or to enhance efficiency. In other words, the spray liquid as applied to the crop is in most cases a mixture or 'cocktail' of ingredients, often specific to particular crop/weed situations (Fig. 4.1). There are close links too between formulation and the method of application, conventionally achieved by nozzles fitted to a boom and fed from a tractor-mounted or trailed reservoir tank. These aspects are especially important at the present time, when a number of novel approaches involving reduction in spray volume and the control of droplet size are under development.

The formulation problem largely centres on the solubility of the active ingredient. Basically, products which are soluble in water are formulated as aqueous concentrates or soluble powders for dissolving, while those soluble only in organic solvents are produced in the form of concentrated emulsions for dilution, a less stable situation. Insoluble materials on the other hand appear as finely ground ('micronised') wettable powders or as partially-formulated suspension concentrates; when added to water these form a relatively unstable suspension of dispersed solid particles. We may consider these in more detail:

(i) *Water-soluble a.i.* From the formulation chemist's and operator's point of view, these are the easiest materials to handle. Examples include many of the basic phenoxyacetic acid and other hormone - type herbicides such as MCPA, 2,4-D, mecoprop and dicamba (usually formulated as their sodium, potassium or amine salts), a variety of contact materials such as ioxynil, paraquat, dinoseb and difenzoquat, the translocated herbicides dalapon and glyphosate and

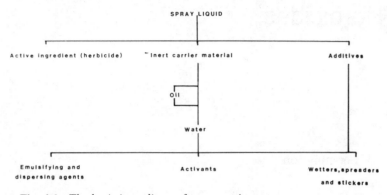

Fig. 4.1: *The basic ingredients of a spray mixture.*

the soil-applied chemical TCA.

(ii) A minority of herbicides which are not water-soluble are formulated as *emulsifiable concentrates* (e.c.), in which the active ingredient is dissolved in organic (petroleum-based) solvents, together with emulsifiers. On adding to water a more or less stable emulsion is formed, requiring only minimum agitation. The nearest familiar equivalent is perhaps cow's milk, and as with the latter there is a tendency to cream formation, which can be minimised by a careful choice of solvent and emulsifying agent. An alternative sometimes used for special purposes (e.g. to reduce spray-drift) is the *invert emulsion*, where the water droplets are dispersed in the organic solvent rather than vice versa. Examples of emulsifiable concentrates include barban, benzolyprop-ethyl, diclofop-methyl, propham, ethofumesate, linuron, pentanochlor and trifluralin.

(iii) *Wettable powders* Very commonly used either where the a.i. is genuinely insoluble in water and in organic solvents or where, for reasons of efficacy or crop safety, this formulation is preferred. Dispersing agents are normally added to offset the natural tendency of these materials to settle out in the spray tank, but some agitation is usually necessary to prevent clogging of spray-lines and nozzles and to ensure that a uniform dose is applied. A clay or silica filler is also often employed to stop these fine particles from caking during formulation or storage.

In general, these formulations have to be mixed in a limited amount of water before final dilution, but this has become easier with the introduction of improved wetting agents and latterly with the development of *suspension concentrates* (flowables, pastes or slurries), which contain the active ingredient and surfactants already dispersed in water. Some of these have been devised for application without further dilution at 2–5 litres/ha via ULV equipment, mainly in non-crop situations. Another alternative for these insoluble materials is formulation as *water-dispersible granules*, which have additional advantages in terms of ease and safety of handling. Examples of each include:

wettable powders (w.p.) – metamitron, metribuzin, chloroxuron, prometryne, carbetamide, propyzamide and lenacil.
suspension concentrates (s.c.) – simazine, terbutryne, propachlor, isoproturon, bifenox, chloridazon mixtures and metoxuron.
water-dispersible granules – cyanazine + MCPA.

4.1.2 GRANULAR FORMULATIONS

As an alternative to water-based types, formulations based on solid carriers have certain advantages; both dusts and granules are used, the former, however, only to a very limited extent. Granules are formed either by impregnating preformed absorptive material with a herbicide solution or by the agglomeration of ready-mixed herbicide and powdered carrier. Advantages claimed for granules include the ability to operate in wet soil conditions and in wind, faster application, no frost damage, easier storage of unused herbicide, absence of drift and greater accuracy and penetration. On the debit side, granules can be costly to produce, are bulky to store and transport, and require specialised application machinery.

A wide range of appliances are employed, ranging from small hand-held shakers to knapsack spreaders and full size tractor-mounted machines. The granules may be dispersed either by gravity or forced-air or by a combination of the two. They are produced to fine tolerances for even flow and should not be mixed with fertiliser or distributed via a fertiliser spreader, which is designed to cope with much greater quantities than are usually required for herbicides.

Established granular formulations include propachlor and tri-allate in arable crops and dichlobenil and chlorthiamid in horticulture, but in view of the advantages mentioned it is not surprising that recent developments have centred on the winter cereal market. A variety of mixtures are available, mainly based on trifluralin, linuron and isoproturon.

4.1.3 ADDITIVES

Reverting to water-based mixtures, still much the most common type, the problem remains (particularly with sprays applied to foliage) of maximising the effect of the herbicide at minimum risk to the crop. In this context interest in the use of surfactants (surface-active agents) has grown considerably in recent years. These materials, generally referred to as wetters, spreaders or stickers, have an effect similar to that of a household detergent, i.e. in reducing the surface tension of the spray liquid. This in turn leads to improved adhesion and spread, so increasing the effective area of contact between herbicide and plant. Another effect of some surfactants, at higher concentrations, is to 'dissolve' the waxy platelets which protect most leaf surfaces, enabling a more thorough adsorption of herbicide and a corresponding improvement in rainfastness.

Aqueous sprays, based on water-soluble active ingredients, can benefit especially from the incorporation of suitable additives, while oil-based and suspension formulations have these qualities to some extent built-in. In many cases appropriate materials are incorporated during manufacture, but where this is not feasible only recommended additives should be used, to prevent excessive run-off or breakdown in formulation stability and selectivity. In some cases, the use of a wetter has become standard practice, e.g. Agral* with difenzoquat against wild oats, and others will no doubt follow.

Finally, the effect of herbicides can occasionally be enhanced by the inclusion of 'activants', which are themselves chemically inert but which have an additive or 'synergistic' effect in combination. These activants or adjuvant oils can be either mineral-based or, increasingly, produced from vegetable sources, including soya beans, sunflower or oilseed rape. Benefits from their use include a smaller range of droplet sizes, reduced evaporation in flight and on the plant surface and, as a back-up to normal wetters, improved leaf-retention and herbicide uptake, especially in sub-optimal spraying conditions.

However, difficulties do arise again in matching the activant to the other spray components, especially wetters. Mismatching can lead to disruption of formulation stability, breakdown in selectivity and increased crop damage. For these reasons PSPS clearances have so far been limited, but include, for example, the use of Actipron* with metamitron and phenmedipham in sugar beet and Cropsafe 11E* with phenmedipham only in beet and with atrazine in maize.

4.1.4 CROP SAFENERS

Whilst valuable, therefore, the use of additives should not be at the risk of jeopardising crop safety. Even if all precautions are taken, the fact remains that the mechanism of crop/weed selectivity is often a tenuous one, dependent not only on correct formulation and application but also on the less controllable factors of weather, crop and weed growth-stage and so on.

For this reason there is a good deal of current interest in crop 'safeners'. These are additives, normally seed-applied, which promote the weed-killing effect of particular herbicides in a context of enhanced crop safety. To date they have been utilised mainly in semi-exotic grain crops, such as maize, especially with the thiocarbamate herbicides (e.g. EPTC), but promising results have also been obtained in perennial ryegrass, giving enhanced selectivity against chickweed, volunteer barley and a variety of annual weed grasses.

The effects of these materials are not yet fully understood, but appear to relate mainly to herbicide breakdown in the crops concerned. Chemicals so far identified include NA (1,8-naphthalic anhydride), cyoxymetrinil, fluarazole and R25788, while others are under scrutiny. They have been reviewed in detail by Stephenson and Ezra (1982), who point to the benefits which might be gained from the parallel development of herbicides and matching safeners for particular crops.

4.1.5 HERBICIDE MIXTURES

Because of the limited range of weeds controlled by individual herbicides, it has become commonplace for manufacturers to formulate them as mixtures wherever possible, but not all permutations can be covered by label recommendations. As a result there has been a marked increase in the practice of 'tank-mixing', both of different herbicides and of herbicides with other agrochemicals. This *ad hoc* mixing (and the use of non-approved chemicals generally) is an especial problem in horticulture, where there is often a lack of guidance. In cereals too, the range of materials applied is considerable and as many as five different chemicals may be mixed in this way.

If products are not fully compatible, however, problems can arise with regard both to their ability to combine in the spray tank and also their resultant performance. It should be noted that compatibility here involves not only the active ingredients, but also any additives which may be present. Mixtures of herbicides with other crop-protection chemicals can pose even greater difficulties, but the advantages of applying more than one chemical at a time are such that manufactuers are under increasing pressure to adopt suitable formulations and a procedure has been agreed with PSPS and ACAS, as a result of which combined recommendations appear on each product label.

4.2 Application

4.2.1 CONVENTIONAL SPRAYING

It has to be admitted that the conventional farm sprayer is a fairly blunt instrument with which to attack often subtle and complex biological problems. Surveys in the late 1970s indicated the extent of the problem in terms of, for example, non-calibration of speed, nozzle operation, and even dosage-rate, and although there is now probably more awareness of the importance of these aspects, the situation still leaves something to be desired (Hudson, 1981).

Crop sprayers have just passed their centenary, the first effective 'modern' machines having been introduced in France and America in 1880 for pest and disease control. Basic sprayer and nozzle design changed little until the introduction of the first low-pressure machines in 1948, following which usage increased rapidly. The spray-tank is normally mounted on the tractor three-point linkage, with a roller-vane or diaphragm pump attached to the p.t.o. and connected to the sprayer unit by suction and delivery hoses. Basic sprayers have a tank capacity of from 200–1000 1, 50-100 1/min. pump output and a 6m-12m boom, although these dimensions are tending to increase, in trailed or self-propelled machines especially.

In horticulture the smaller mounted machines are often more appropriate, while considerable use is also made of hand-operated knapsack sprayers, especially in parks and gardens. At the same time a number of more or less specialised methods have been developed to deal with container-grown plants, soft fruit and vegetables (q.v.). In orchards, centrally-mounted booms are unsuitable and have been replaced by high-pressure 'one-sided' nozzles such as the Spraying Systems TeeJet OC*. Various forms of guarded or directed sprayers are also used, especially in intensive row-crop production and in strawberries.

It is not the intention of this text to provide a comprehensive treatise on sprayer design and operation, which is well covered elsewhere (e.g. Roberts, 1982). Attention is drawn, however, to a number of points of significant practical importance, such as the need for at least some agitation of wettable-powder and emulsion-type formulations; this is carried out on all but the largest sprayers by means of recirculation of the spray liquid. Boom height and stability are also important, especially where fan nozzles are used. Long booms are particularly subject to both lateral and vertical movement and much ingenuity has recently gone into the production of booms with improved rigidity characteristics and other significant improvements, such as universal link suspension (Nation, 1982).

Another important aspect is that of nozzle-type. Two main kinds are employed; cone (or 'swirl') nozzles, which are now mainly used to produce a coarse spray for aerial and other low-drift situations, and fan nozzles, which give a sheet of spray, either elliptical or rectangular in outline (Fig. 4.2), and are the basis of most field operations. Problems have arisen over nozzle terminology, and a standard system of coding proposed by Gardner (1983) appears to have much to commend it. In this system, for example, F110/1.1-1.2/3.0 would represent a fan nozzle operating over a 110° angle, with a capacity of 1.1 to 1.2 litres/min. at a pressure of 3.0 bars.

Whatever the type of nozzle, it is the final spray distribution relative

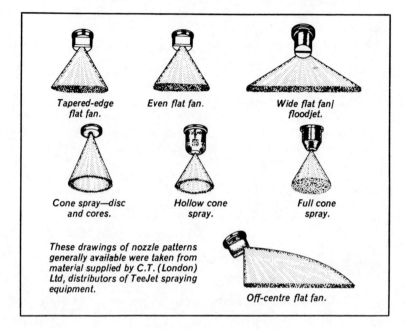

Fig. 4.2: *Spray nozzle patterns (After Marshall, 1981).*

to the target which matters. The preferred pattern is an overlapping one, such that the outer edge of one triangular fan is below the centre of the next, but this will only be achieved if boom-height is correctly set (Fig. 4.3). Calibration of individual nozzles can easily be carried out with an electronic flow meter or other monitoring device and, given the very variable nozzle performance recorded in on-farm surveys, this would appear to be one routine task that is well worth carrying out, when the cost of replacement nozzles is minimal compared to that of the whole operation.

4.2.2 LOW-VOLUME APPLICATIONS

Concern has grown recently over two aspects of conventional spraying. These are, firstly, the amount of water-carry involved (up to 400 l or more per ha) and, more fundamentally, the effects of the variable droplet sizes produced. In this connection it may be noted that various categorisations of spray volume and droplet-size are possible, one example being shown in Table 4.1.

Droplet size is normally expressed in terms of volume median diameter (vmd), measured in microns*; this is to say that in a given

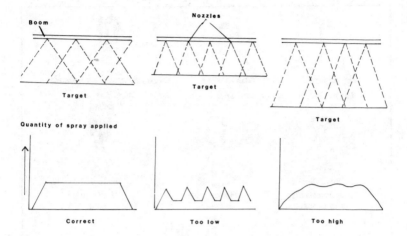

Fig. 4.3: *The effect of boom height on spray deposition pattern. (after Roberts, 1982).*

Table 4.1 *Classification of sprays by droplet size and volume application rate – herbicides*

A. *By droplet size*

Description	Range of v.m.d.[a] (μm)
Coarse spray	500
Medium spray	200–500
Fine spray	100–200
Very fine spray/mist	30–100
Aerosol	30

[a]Volume median diameter

B. *By volume application rate*

Description	Application rate (l/ha)	
	Field crops	Bushes and trees
High volume (HV)	600	1000
Medium volume (MV)	200–600	300–1000
Low volume (LV)	50–200	50–300
Very low volume (VLV)	5–50	20–50
Ultra-low volume (ULV)[b]	5	20

[b]Essentially waterless formulation

(*After Johnstone, 1978*)

sample of spray, half the volume will be composed of droplets larger than the vmd and half of smaller droplets.

(* 1 micron (μm) = $\frac{1}{1000}$ of a millimetre.)

Another frequently used measure is the number median diameter (nmd); in a given spray liquid, half of all the droplets will be larger than the nmd and half smaller. These two measures give prominence to large and to small droplets respectively; the ratio between them (vmd/nmd) therefore reflects the range of droplet sizes in the spray. The closer the ratio is to unity the more uniform the spray.

Anxiety over water-volumes must be seen against the background of increasing use of herbicides in the autumn and winter, when deteriorating soil conditions can make access difficult. One main spin-off of this has been the development of a wide range of low ground pressure (l.g.p.) vehicles such as the Argocat* and Highland Garron* which, by means of multi-axles and/or low-pressure 'flotation' tyres, provide increased flexibility on wet or heavy land (Rutherford, 1980). Developed partly at the Weed Research Organisation, these vehicles have been widely adopted in the more intensive cereal-growing areas.

That apart, however, the first commercial application of reduced volume spraying dates back to the mid 1970s when Ciba-Geigy introduced their '7-gallon system' for use with the soil-applied herbicides isoproturon, chlortoluron and terbutryne. This involved spraying at around 80l/ha (as against the normal 200-400l/ha), utilising a range of specialised low-pressure nozzles. Other manufacturers followed suit and similar low volume systems are increasingly being adopted in a range of crops, with or without the use of l.g.p. equipment. The advantages can be summarised as follows:

(i) Less water required.
(ii) Reduced filling-time and fewer trips to the water-point or tanker.
(iii) More spraying time, therefore a greater area sprayed per tank or per day.
(iv) Perhaps most importantly, improved timeliness of spraying, allowing the herbicide to be applied to the best advantage relative to soil conditions and to crop and weed growth-stage.

4.2.3 VARIABILITY IN DROPLET SIZE – HERBICIDE DRIFT

The problem here can be simply stated: with conventional, especially high volume applications, a range of spray droplets is produced, the majority of which are largely ineffective in their main object of killing weeds. Those under about 100 microns v.m.d. are so light as to be increasingly unaffected by gravity and are therefore likely to drift off without coming into contact with the target. They are, however, capable of causing severe problems in nearby susceptible crops or amenity/wildlife areas. Conversely those with a v.m.d. of >350 microns, because of their relatively great momentum, tend to bounce off the target, again without having any effect. It is worth bearing in mind too, that, because of their greater volume, these large droplets actually contain the bulk of the herbicide.

Within the limits of the conventional system, variability of droplet size has been manipulated so as to serve a variety of weed control purposes. A distinction can be drawn, for example, between hormone weedkillers, in which a relatively coarse spray is less likely to cause drift problems, and contact types, whose effect is dependent on maximum coverage of weed foliage, best achieved by fine droplets.

The problem of herbicide drift has been increasingly recognised in recent years, partly as a result of the spread of susceptible crops such as oil-seed rape and has been extensively reviewed in a recent B.C.P.C. publication (Elliott and Wilson (eds.), 1983). There are three ways in which it can take place:

(i) *Spray-drift* This occurs when small droplets of spray liquid are carried away from the target area by the wind.

(ii) *Vapour-drift* In this case, vapour from volatile herbicides is carried away from the treated area at or after the time of spraying. This is again caused by wind, or by warm air currents.

(iii) *Blow-off* This type of drift is normally associated with light soils, particularly in situations where there are no hedges. It is due to the movement, in windy conditions, of tiny particles of treated soil.

Of the three types spray drift is the most common. While this may on occasion cause damage to adjacent agricultural crops, the majority of damage, and the most serious, occurs in horticulture. This is often the result of spraying grassland or cereals with growth regulator-type herbicides such as 2,4-D and mecoprop, to which a large number of outdoor and glass-house-grown horticultural crops are susceptible, although even contact herbicides may damage the foliage of ornamental crops, so reducing their value. If users could make themselves aware of nearby horticultural establishments and inform growers of spraying schedules, much damage could be avoided.

In any event a number of precautions can be taken to ensure that damage is minimised, e.g. by arranging spraying to coincide with periods of calm weather. Low pressure nozzles should be used, as these tend to give a greater proportion of large spray droplets and, in addition, a slight lowering of the spray boom in combination with adjusting the nozzles backwards at an angle of about 40° to the vertical will reduce the risk of drift. When growth-regulator herbicides are used, spraying should be carried out, if possible, before adjacent susceptible crops have emerged and it may also be worthwhile to leave an untreated strip of one or two boom widths alongside such crops; weeds in this area may be treated with a contact herbicide at the beginning of the growing season.

Vapour drift has hitherto been considered of little significance in the relatively cool British summers, but recent trials with ester formulations of mecoprop and dichlorprop have revealed damage in weather conditions ranging from fine and warm to cloudy and very cool, indicating that the problem may be greater than imagined (Eagle, 1982). Damage to test plants was recorded outside the treated area up to 30 hours after spraying, at a distance of 50m. in cool weather and 100m. in fine warm weather.

Although chemical companies have attempted to overcome this problem by introducing amine formulations with a reduced tendency to vaporise, the number of product labels which provide adequate warning and advice on vapour drift is still too low. If possible, therefore, volatile formulations of growth regulator herbicides should not be applied in high risk situations, especially in warm still conditions. Attempts have been made to quantify both spray- and vapour-drift and related meteorological factors with a view to improving the predictability of drift-risk and of suitable spraying conditions, but more work will be required to clarify this complex situation (Thompson, 1982).

The fact remains, however, that no matter which system is employed there will inevitably be a proportion of small, drift-prone droplets in the spray. One solution to this has been through the use of 'anti-drift' agents such as Driftless* and Admongel*. These are polyacrylamide gel substances which cut down the number of small droplets by increasing the viscosity of the spray liquid or by reducing the 'energy-drop' at the spray-nozzle. Their main advantage in practice appears to relate to the ability to allow spraying to continue in windy conditions, although it may be noted that comparable results can be obtained simply by reducing pressure.

*[The symptoms of damage caused by herbicide drift and from other sources, such as over-dosing, leaching, contamination, soil residues and wrong timing, are fully described and illustrated in colour in MAFF Reference Book 221, *Diagnosis of Herbicide Damage to Crops*, HMSO, 1981.]

4.3 New Application Methods

4.3.1 CONTROLLED DROPLET APPLICATION (CDA)

A more revolutionary approach to droplet size lies in the concept of controlled droplet application (CDA), one of the most important developments in pesticide technology in recent years (Matthews, 1979). In essence the technique involves the production of a restricted range of droplet sizes, by means of a novel system in which droplets are generated by feeding the spray-liquid on to a rapidly-rotating, serrated-edged disc or cup.

The centrifugal force produced by such 'rotary atomisers' causes the liquid to be thrown off in the form of more or less equal-sized drops; this compares with the conventional system, where a sheet or cone of liquid breaks up in an uncontrolled way into droplets of greatly varying sizes (Fig. 4.4). Droplet size can be controlled within fairly narrow limits by varying either the speed of rotation (r.p.m.) of the disc or the flow-rate of the spray-liquid, higher r.p.m. and lower flow-rates producing smaller drops. The maximum flow-rate (about 80l/ha) is governed by the need to prevent the droplet-producing ligaments from coalescing.

The absence of wasteful large droplets allows substantial reductions in the volume of liquid applied, while the elimination of drift and evaporation-prone microdroplets and the ability to tailor droplet size to target requirements are other widely aired potential advantages (e.g. Bals, 1982). The technique therefore aroused great interest when it was introduced in the mid 1970s, but difficulties in translating its potential into field performance led to a period of some stagnation in its development. Recent design changes, however, including improved boom stability, the use of hydraulic instead of electrically-powered atomisers and 'in-cab' selection of droplet-size have renewed interest in the concept.

In general, results from CDA treatments have been variable and with contact herbicides at least, somewhat inferior to those obtained by conventional means; of all the herbicides tested only glyphosate has shown a consistently improved performance (Taylor, 1981). Recent evidence suggests, however, that improved control may be obtained with droplets of around 150-175μm and a water-volume of around 40 l/ha, as against the 20 l/ha sometimes employed (Bailey et al, 1982).

One particular point at issue relates to the feasibility of reducing herbicide dosage rates. CDA enthusiasts take the view that with fully integrated systems significant reductions can be achieved, with no falling-off or even with enhanced results. In general, however, both commercial and trials results tend to cast doubt on this, and wider opinion suggests that CDA is likely to be fully competitive only in near-optimal conditions and that in most field situations conventional systems will have the edge.

4.3.2 ULTRA-LOW-VOLUME SPRAYING (ULV)

Another extension of the above approach is that of ultra-low-volume spraying or ULV. The term is sometimes used as if it were synonymous with CDA but this is misleading. Both involve the production of more or less uniform droplets, but while CDA may be obtained over a range of spray volumes, ULV relates specifically to the use of extremely low

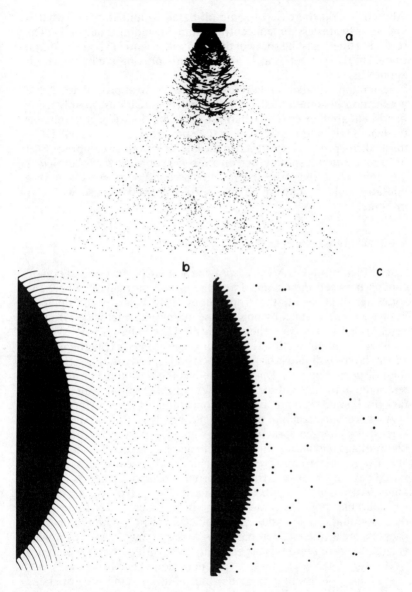

Fig. 4.4: *Droplet formation from; (a) a conventional pressure nozzle (b) a spinning disc (c) a spinning disc with a serrated edge (after Roberts, 1982).*

dilutions or, in the case of some oil-based formulations, to what is known as 'waterless' or 'concentrate' spraying (Johnstone, 1978). The result is a fine mist-like spray of droplets with a vmd of about 50-70μm, carried to their target by air-currents ('drift-spraying') or by forced-air apparatus.

A limited number of herbicides have been approved for ULV application in some grassland and non-crop situations, mainly using hand-held appliances (e.g. Micron Ulva*), but its wider role remains unclear. One field application is the somewhat controversial Ulva-mast* drift-sprayer, introduced in the late 1970s, which comprises a set of 15 rotary atomisers mounted on a vertical boom. So far as herbicides are concerned the drift danger to susceptible crops is a severe limitation and to date none has been approved for use with this machine.

4.3.3 ELECTROSTATIC SPRAYING

One difficulty of CDA/ULV spraying is that of ensuring adequate contact between the chemical and the target, particularly where the spray has to penetrate the crop canopy. Efforts have been made to improve on this aspect by angling the atomisers (Microcide Ground-plane*) or by arranging them vertically (Tecnoma Girojet*), while another recent innovation is the rotary cage atomiser (Micronair*), in which the spray-liquid is broken up by a high-speed rotating screen and then dispersed by fan. This machine has the advantage of not being restricted to the level of spray volumes typical of spinning discs, but so far only limited data are available concerning its field performance.

A more far-reaching development involves the production of electrically charged spray droplets, which are positively attracted to plant foliage and at the same time repelled by one another, leading to enhanced contact and distribution (Fig. 4.5). Because drift is mini-mised (by this positive attraction) droplet size and hence total spray volume can again be significantly reduced. The system appears to offer considerable potential, but results in practice have again been disappointing, possibly due to the efficiency with which the charged droplets are attracted to crop foliage, so protecting the lower-growing weeds. Research into this aspect continues, with particular reference to the use of aerofoils to improve penetration (Lake et al. 1982).

A number of different methods of forming, charging and dispersing the droplets have been tested, including the ICI Electrodyn*, initially developed as a hand-held machine, although tractor-mounted versions are now under development. Work is also in hand at the National Institute of Agricultural Engineering to develop a system to charge conventionally produced droplets, which may circumvent some of the penetration problems.

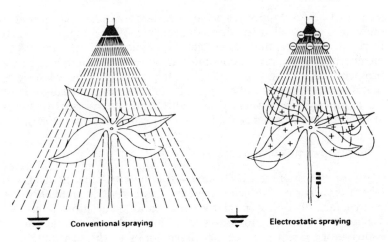

Conventional spraying   Electrostatic spraying

Fig. 4.5: *The wrap-around effect of charged droplets (after Chambers, 1981).*

4.3.4 SELECTIVE APPLICATION (SELAP)

A quite different method of applying herbicides which has had a rapid acceptance in the U.S.A. and more recently in Britain is that involving the use of applicators to selectively control tall-growing weeds. The technique has a considerable potential in low-growing crops such as strawberries, carrots and peas; against docks, nettles, rushes and thistles in grassland; and against certain arable weeds, including wild oats, weed beet and volunteer potatoes.

Three main types of applicator have been developed – recirculating sprayers, rollers and carpet-applicators, and rope-wick machines; the most successful of these are the rope-wick types, invented in America by J. E. Dale in the late 1970s. In these the herbicide, usually glyphosate, is smeared on to the weeds by means of saturated 'wicks' attached to a tubular reservoir. The machines are basically simple and robust and both front and rear tractor-mountings are possible, the latter preventing the tractor from running over treated weeds. They include the Hectaspan Weed-Wiper,* which has horizontal wicks, the Tecnoma Topweeder* and Keenan Weed-licker*, with vertical, overlapping wicks and the Vicon Wedge-Wik* in which the wicks are held in a horizontal 'V' shape, intended to enhance contact with the weeds. A variety of hand-held appliances for horticultural and general round-the-farm use are also available.

These rope-wick machines are generally less expensive to buy and maintain than other selective applicators. Early problems over the uneven distribution of herbicide on sloping ground and intermittent feed to the wicks appear to have been largely overcome, although difficulties can still arise through clogging of the ropes by weed fragments. The fact that the herbicide is applied directly to the weeds means that considerable economies are possible, although practical problems exist, notably in sugar beet. In grassland, evidence suggests that a height difference of between 10 and 20cm is sufficient, at least where the crop is relatively uniform, although in many cases two passes at right-angles may be necessary for best results (Lutman, 1980).

4.3.5 ELECTRONIC CONTROL SYSTEMS

Many of the discrepancies which are apparent between field and laboratory data on herbicide effectiveness can be attributed to the relative lack of control in the former situation. Variations in ground surface, soil condition, forward speed, spray volume and other minor contributory factors can lead to fluctuations in the distribution of active ingredient over the target area.

Systems are now available, however, which allow in-cab monitoring of various aspects of the spraying operation and which in some cases automatically take appropriate corrective measures (e.g. Agmet Automatic Pacemaker*; R.D.S. Auto Volume Regulator*). In general these operate by adjusting the spray pressure at the nozzles in response to tractor speed and flow rate; this has the disadvantage, however, of altering droplet size and hence physical and biological performance. One solution may be to separate the liquid carrier and the active ingredient, so that the former is fed from the tank to the boom at a constant rate while the latter is metered in from a separate reservoir at a rate commensurate with the forward speed of the vehicle. This would ensure consistency both of droplet size and of dosage rate of a.i. per unit area.

Possibilities exist for extending this electronic monitoring and control to a number of other aspects of spraying, including boom stability and swathe-matching, again in the interests of tightening up the whole operation. Control of boom movement both vertically (= 'bounce') and horizontally (= 'whip' or 'yaw') is likely to be achieved mainly by a switch from passive spring and damper suspensions to active systems utilising height sensors to maintain boom stability within narrow limits. By enabling the boom to operate closer to the ground this could improve penetration, reduce drift (especially in marginal spraying conditions) and contribute generally to a more even spray distribution.

Accurate swathe-matching is also becoming increasingly important. Normally carried out by means of headland markers, trailing cords from the boom ends or tramlines set up at drilling, this aspect still leaves something to be desired. The use of foam-blob markers, with or without the aid of mirrors and/or electronic sensors is one possible answer, while in the longer term more sophisticated navigational techniques based on marine or aerial systems may have a part to play, if technical and cost problems can be overcome (Lawrence, 1980).

4.3.6 RÉSUMÉ

This is a period of considerable upheaval in the field of herbicide formulation and application. Intensification and specialisation of cropping, pressure on variable costs and an increased awareness of environmental issues have combined to foster interest in ways of reducing chemical wastage and improving overall spraying efficiency. This is particularly true of cereals, where current yield levels are heavily dependent on the use of crop protection chemicals, especially herbicides. Much impetus has come from the WRO, where, for example, the combination of reduced water volumes and fast low ground pressure vehicles has been seen to offer major advantages, especially in the adverse spraying conditions often associated with winter cropping (e.g. Ayres, 1980). These methods are contrasted with conventional spraying, with its extra water-carry and relatively inefficient utilisation of active ingredient.

The fact remains, however, that in most situations the conventional system does work and indeed, with current improvements in design and a generally more flexible attitude to volume and pressure rates, its position has if anything been strengthened. At the same time the use of charged droplets may open the way to improved results at existing dosage rates or, perhaps more importantly, to equivalent results at reduced rates, with benefits both in terms of cost and of crop safety.

The development of 'closed' spraying systems will also improve the accuracy and safety of herbicide applications. These systems, at their most sophisticated, involve the provision of sealed 'plug-in' containers, which dispense with the sometimes irksome and potentially hazardous process of mixing the herbicide. Equally they avoid the need to dispose of empty containers or surplus chemical, environmentally the most risky part of the whole operation. Meanwhile, fundamental research goes on into improving our understanding of the interaction between weeds and herbicides. For example, recent work indicates that the effect of some herbicides and wetters can be significantly enhanced by localised application to specific leaves or other parts of the plant (Merritt, 1980). It remains to be seen whether application technology will be able to meet this particular challenge as successfully as it has others.

Further Reading
Various authors, *Spraying Systems for the 1980s*, British Crop Protection Council Monograph No. 24, 1980.

Useful collection of papers, covering various aspects of spraying technology. Main papers included in Selected Bibliography.

ELLIOTT, J. G. and WILSON, B. J. (eds.) (1983), 'The Influence of the Weather on the Efficiency and Safety of Pesticide Applications: the Drift of Herbicides', *Occ. Pub. No. 3*. Br. Crop. Prot. Council, Malvern.

Definitive compilation of information on this important topic.

HUDSON, J. E. (1981), 'Application of pesticides by conventional equipment', *Proc. Crop Prot. in Northern Britain Conf., 1981*, pp. 169-81.

Concise review of background to newer developments.

NATION, H. J. (1982), 'Application Technology: Review and Prospects', *Proc. 1982 Br. Crop Conf. – Weeds*, pp. 983-94.

Excellent review of new application techniques. Extensive bibliography.

ROBERTS, H. A. (ed.) (1982), 'The Application of Herbicides' *Weed Control Handbook*, 7th edn., Blackwell Scientific Publications, Oxford, pp. 158-218.

Background reading, covering all aspects of conventional spraying and dealing briefly with recent trends.

Various authors, *Outlook on Agriculture*, *10*, 1981.

Collection of papers dealing with new developments in herbicide application. Valuable insight into international background of developments.

5 Weed control in cereals

5.1 The effects of weeds in cereals

One of the main difficulties in assessing the effects of weeds in cereals is the variety of ways in which these effects are expressed. Most obviously there is the matter of direct competition, leading to the possibility of reduced yields. However, other factors such as grain quality, ease of harvesting, and disease carry-over may be at least as significant in many cases. Most investigations in this area have involved both broad-leaved and grass weeds but only wild oats and blackgrass have received a significant measure of individual attention; additional material in respect of these is presented in the relevant sections. General reviews include those of Elliott (1978; 1980) covering mainly harvest and post-harvest effects and Snaydon (1982), covering crop/weed competition, with particular reference to cereals.

5.1.1 COMPETITION WITH THE CROP

In general terms the severity of this will tend to increase with weed numbers, but, as indicated, there have been relatively few attempts to quantify its effects, other than in the major grass weeds (q.v.). In some cases linear relationships have been reported between weed density and yield reduction (e.g. Wells, 1979), but in others a 'curvilinear' response has been found, such that at higher weed densities the effect on crop yield begins to level off (e.g. Scragg, 1980). Most assessments have involved yield responses to the removal of weeds by herbicides, but often without the weed infestation level being specified. More penetrating studies, incorporating the different parameters of yield (ear number, grain number, etc.), are, however, increasingly being reported (e.g. Courtney and Johnston, 1982).

Grass weeds, which parallel the cereal in morphology (both above and below ground) and in physiology, might be expected to compete more effectively, and this indeed appears to be the case. Baldwin (1979), for example, gives mean yield responses to wild oat and blackgrass removal of 10-20%; a series of similar trials involving broad-leaved weeds produced a corresponding figure of 2% (Snaydon, 1982). The effects of broad-leaved weeds are also less predictable. The obvious competitors include large species such as poppies, charlock, hemp-nettle and mayweeds, but lower-growing types such as field pansy, chickweed and speedwells can also affect yields.

Competition in the early stages is mainly for mineral nutrients (especially nitrogen) and for moisture, and appears to be generally more damaging than the shading effects which can occur later in the life of the crop. The interaction depends very largely, however, on the relative growth rates of the crop and the weeds and anything which reduces crop vigour is undesirable. Factors such as nutrient deficiency, low pH, poor drainage, soil compaction, presence of disease, and too-deep sowing can all predispose the crop to weed attack and their importance should not be underestimated.

Crop density too can be an important factor in combating weeds, particularly in the early stages of growth, and one suggestion is that where a high level of weed competition is expected, appropriate levels of seeding should be used to ensure a cereal plant population of at least 250 seedlings/m^2 (Roberts, 1982).

5.1.2 EFFECT OF WEEDS AT HARVEST

With the almost universal adoption of combine harvesting, the problems caused by weeds at this stage have been increasingly recognised. In general terms weeds make the process more time-consuming, increase grain loss and contaminate the grain saved. Green weeds in the base of the crop hold moisture, so shortening the combining day, while climbing and clinging weeds such as cleavers and the bindweeds act as a focus for lodging.

The main effect is to increase the amount of material entering the combine (= throughput), resulting in incomplete separation of the grain and/or reduced forward speed. If combine speed is maintained in a weedy area there is an increase in grain loss, which escalates rapidly over a certain level (Fig. 5.1). Alternatively, if speed is reduced grain losses can be minimised, but with a corresponding time penalty. Elliott (1980), in reviewing this topic, stressed the importance of keeping matter other than grain, or MOG, below the level at which grain losses are likely to escalate at constant speed and pointed out that in a weed-free situation the grain/MOG ratio is often about unity, i.e. roughly equal amounts of grain and non-grain material are present. The presence of weeds can reduce this ratio by varying amounts and on this basis Elliott identified three main weed categories:

  (i) wild oats, blackgrass and many annuals – these compete during the life of the crop, reducing the grain yield, but are largely dessicated by harvest time.

  (ii) late-emerging, mainly perennial weeds such as field bindweed, sow-thistle and creeping thistle, which are unlikely to affect yield but may have a significant effect at harvest.

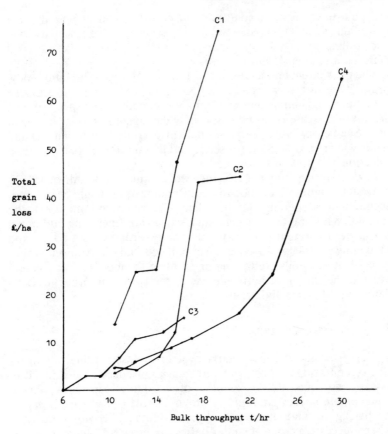

Fig. 5.1: *The effect of throughput on grain loss in four combine harvesters (after Elliott, 1978).*

(iii) weeds which compete in the growing crop and remain green at harvest, so causing both types of loss – these include, for example, couch, cleavers, mayweeds, corn marigold and chickweed.

5.1.3 POST-HARVEST EFFECTS

The presence of weed seeds and other fragments is undesirable at any time and especially so in specialist crops destined for milling, malting or the seed-trade itself. EEC intervention standards permit no more than 3% total impurities in feed grain, and with this outlet assuming increasing importance, prevention of seeding must warrant careful attention. Grain which exceeds this limit has to be cleaned before acceptance, which can lead to a double penalty involving both the cost of cleaning itself and a loss of weight adjustment. In the case of malting and milling grain too, premiums may be forfeited if excess weed material is present (Elliott, 1978).

The most direct impact, however, is on seed crops. Regulations now prohibit the offering for sale of seed grain which has more than a prescribed maximum content of total weed seeds, while specific limits are set in the case of certain especially damaging weeds such as wild oats. Seed crops must also pass a field inspection prior to harvest and fields which are intended for seed production require special care in selection and previous cropping.

If wild oat is present, herbicides which stunt the weed, making it difficult to rogue, should be avoided. Roguing itself should be by hand, as the chemical roguing-glove may leave dead seeds to contaminate the sample. With grass weeds generally, extra care (including additional herbicide treatments) may be necessary around field-edges and ploughing is to be preferred as the initial seed-bed preparation. Where broad-leaved weeds are the major problem the main concern should be with the timing of herbicide use in the spring, in order to avoid damage to the ears and grains.

5.1.4 DISEASE CARRY-OVER

Losses associated with the carry-over of disease-causing organisms (the 'green-bridge' effect) are widely acknowledged, but difficult to assess and quantify. Examples often quoted include take-all (*Gaeumamnomyces graminis*) on couch, ergot (*Claviceps purpurea*) on blackgrass and, more recently, net-blotch (*Pyrenophora teres*) where successive crops of winter barley are grown. Volunteer cereals themselves are almost certainly the worst culprits and Elliott (1978) provides specific examples, including one involving yellow rust (*Puccinia striiformis*), where a yield loss at 1978 prices of around £70/ha was attributed to volunteer wheat in a winter wheat crop.

5.2 Broad-leaved weeds in cereals

5.2.1 ANNUAL BROAD-LEAVED WEEDS

There appears to be a general consensus that common chickweed is the most widespread and abundant weed species in both spring and autumn-sown cereals (e.g. Makepeace, 1982, b). In the former it is accompanied most often by the 'polygonums', charlock and fat hen, while in winter crops there is a predictably high incidence of predominantly autumn-germinating species, such as cleavers, may-weeds and speedwells (Table 5.1).

Table 5.1: *Broad-leaved weeds infesting cereals in the UK – a combination of light and heavy infestations.*

| Weeds | % of acreage infested | |
	Spring cereals	Winter cereals
Chickweed	83	89
Knotgrass	69	42
Black bindweed	64	25
Charlock	58	33
Redshank	50	19
Fat hen	50	13
Mayweeds	44	53
Speedwell	33	55
Hemp-nettle	28	12
Cleavers	22	52
Fumitory	20	12
Thistles*	19	16
Spurrey	17	4
Poppy	14	20
Docks	14	12
Field pansy	13	14
Shepherd's purse	13	11
Groundsel	11	17
Annual nettle	11	5
Parsley piert	4	16

*Including sow-thistles.
(*After O'Leary, 1973*).

Within this overall framework, a good deal of regional variation exists, as we have seen. In Scotland, for example, redshank, spurrey and hemp-nettle are common in spring crops, while in winter cereals, although cleavers and speedwells are relatively infrequent, species such as shepherd's purse, red dead-nettle and forget-me-not could cause problems as the area sown to these crops increases.

5.2.2 PERENNIAL BROAD-LEAVED WEEDS

Perennial species such as docks, thistles and field bindweed may arise from seed or from root, rhizome or other fragments. Many, including docks and creeping buttercup, are readily controlled as seedlings along with the annual weeds. Otherwise the maximum level of control may not be available until the weeds have developed a reasonable amount of foliage, which is likely to be later than the optimum time for spraying annuals. In some cases too, the most effective chemicals are unsuitable for application to cereal crops and control has conventionally been carried out in the stubble. Makepeace (1982, b has referred to the possibility of an increase in species such as hogweed, corn mint and hedge woundwort in minimum tillage situations, although in general perennials are less tolerant of arable conditions.

5.2.3 SPRING CEREALS

Control of broad-leaved weeds is more straightforward in spring cereals than in winter crops. Germination of most species takes place over a relatively short period, during late April in the south and early May in the north, so that it is possible to control most annual weeds most of the time with one spray application. Of the major species, knotgrass and fumitory tend to be somewhat earlier than this and redshank and fat hen somewhat later.

The choice of chemical is also easier for spring cereals, the great majority of treatments being at the post-emergence stage. The herbicides involved include a wide range of single chemicals or mixtures, the latter being generally more expensive, but controlling a greater variety of weeds. Translocated chemicals (MCPA; 2,4-D; mecoprop, etc.) form the main line of attack, while the contact types (ioxynil; bromoxynil, etc.) are also effective and generally less damaging to the crop. Many proprietary mixtures now contain both types, thus achieving the twin aims of wide-ranging weed control and adequate crop safety.

In control terms the weeds range from a basic MCPA-susceptible group (hemp-nettle, charlock, fat hen, etc.), through resistant types such as chickweed, knotgrass and mayweeds to the more intractable speedwells and corn marigold. It is fair to say, however, that most combinations of broad-leaved weeds in spring cereals can be adequately controlled by one or other of the range of herbicides and mixtures presently available. Full details of these, covering dosage rates, cost, range of weeds controlled and any special considerations are given in MAFF Booklet 2252, revised annually.

Two further points may usefully be dealt with at this stage. Firstly, apart from these post-emergence treatments a few materials are now available for pre-emergence use, including both liquid and, more recently, granular formulations. The scope for such residual materials is less, however, in spring than winter cereals, owing to the greater risk of damage to following crops. Secondly, it should be noted that many of the above herbicides are only suitable for cereals not undersown with clover. Where clover is present the choice of chemical is restricted to certain of the contact types, including ioxynil/bromoxynil and benazolin, and to the butyric acid formulations MCPB and 2,4-DB.

5.2.4 SPRING CEREALS – TIMING OF APPLICATION

The fact that most cereal herbicides are applied post-emergence means that herbicide tolerance is a major consideration, in both spring and winter crops. The basic mechanisms of selectivity are discussed in chapter 3; it suffices to note here that, in cereals, as in other crops, these can be adversely affected, for example by over-dosing or other forms of misapplication, by weather conditions around the time of spraying and by the condition of the crop itself. It should also be remembered, in this connection, that the safety of a herbicide mixture is only as great as that of its least safe component and that in general herbicides should only be applied in mixture with other agrochemicals where clear guidelines exist as to their compatibility.

The critical factor here, however, is the stage of growth of the cereal plant at the time of spraying. For weed control purposes this is usually expressed in terms of the number of fully expanded leaves on the main shoot (i.e. excluding tillers). In spring cereals especially this bears a sufficiently close relationship to ear development to enable its use as a reliable field index in most situations (Fig. 5.2). In spring crops, therefore, spraying with translocated ('growth-regulator') herbicides should normally take place from the 'five-leaf' stage up to, but not including, the appearance of the first node at the base of the stem – the 'jointing' stage or G.S. 30 in the terminology of Zadoks et al. (1974). (For a full explanation of the Decimal Code for the Growth-stages of Cereals, see Tottman, D. R., and Makepeace, R. J. (1982), B.C.P.C. Occ. Pub. No. 2.). With mainly contact-acting herbicides or mixtures a somewhat wider range of timings is possible, but it should be emphasised that, in all cases, spraying outside the recommended growth stages leads to a real possibility of crop damage (Tottman, 1982).

The symptoms of such damage are well-known (Roberts 1982). Both leaf and ear primordia can be affected by the application of herbicides at the early differentiation stage of development, i.e. before the primordia actually become visible at the apex. Damage is usually expressed in the form of tubular or 'onion' leaves (which can trap the emerging ear), enlarged and fused floral parts and gross ear deformities, including both disrupted and additional spikelets. In barley especially, ears may be weakened and eventually break off and grain quality may be reduced. There is some evidence, however, that the overall effects on grain yield may be more severe in wheat.

Information relevant to the end of the safe stage is more scarce and relates mainly to the benzoic acids dicamba and 2,3,6-TBA. Late application of these materials typically leads to the development of 'rat-tailed' ears, with thin, shrivelled grain, and treatment once the first node is detectable is likely to have serious effects on subsequent yield (Tottman and Makepeace 1982).

In general terms, therefore, spring cereals should be treated at the

(a) One fully expanded leaf,

(b) Three fully expanded leav.s,
main shoot and one tille:.

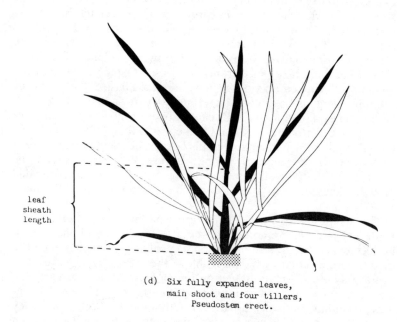

leaf
sheath
length

(d) Six fully expanded leaves,
main shoot and four tillers,
Pseudostem erect.

Fig. 5.2: *Growth stages in winter wheat (after Tottman, 1976).*

Cereal plants drawn
by Hilary Broad.

(c) Five fully expanded leaves,
main shoot and three tillers.

(e) Eight fully expanded leaves,
main shoot and three tillers,
Two nodes detectable.

earliest opportunity commensurate with crop safety. Weeds are more susceptible to herbicides at younger growth stages, while early removal also reduces competition. The length of time available for spraying in the spring is, however, limited. Modern cereal varieties can pass through the critical growth-stages in a fortnight or less and this narrow 'spraying window' may be further reduced or even eliminated by other constraints. For example, a reasonable rain-free period is necessary for the contact herbicides particularly, while on the other hand the translocated types are liable to cause drift problems in windy conditions, especially in the vicinity of glasshouses and susceptible crops such as oil-seed rape. Another restriction relates to the need for a gap of up to ten days between the use of broad-leaved and some wild oat herbicides, which may disrupt the timing of both.

5.2.5 WINTER CEREALS

We have already referred to the distinctive weed flora of winter cereals, which poses new problems of control. The weeds concerned include a number (e.g. chickweed) in which germination is instigated by the fall in soil temperatures in September, later types such as cleavers and ivy-leaved speedwell, and others, such as field pansy, which germinate both in autumn and in spring. There can, of course, also be competition from spring-germinating annuals.

The range of herbicides available was originally confined to the standard contact and translocated types, applied in the spring. A breakthrough came with the realisation that the major blackgrass herbicides also controlled dicot. weeds, when applied at reduced rates in the autumn. Interest in this approach intensified following reports of significant yield advantages from the early removal of these weeds and, stimulated by this and by the growth of the winter cereal sector, there was a rapid increase in the range of products, including various combinations of isoproturon, chlortoluron, linuron, trifluralin and other soil-applied, mainly pre-emergence chemicals, in both liquid and granular formulations (Cussans, 1980).

Considerable success was achieved with these, but their intensive use resulted in a build-up of resistant species, notably cleavers and some speedwells. This led in turn to renewed interest in early post-emergence application of mixtures of mecoprop, cyanazine and MCPA with contact types such as bromoxynil and bronofenoxim and also to the rapid uptake of new soil-applied chemicals such as pendimethalin, bifenox and chlorsulfuron.

Cereal workers have also responded to developments in other crops by examining the effectiveness of low-dose, sequential treatments of mecoprop/contact mixtures. Promising results have been achieved, based on two applications in autumn/winter and in spring (e.g. Bradford and Smith, 1982). These treatments reduce costs and allow a greater margin of crop safety and hence flexibility. They therefore have potential either on their own or as a follow-up to a pre-or early post-emergence herbicide, where the more difficult weeds are involved. Full details of chemicals etc. are again given in MAFF Bulletins 2252 and 2253.

5.2.6 WINTER CEREALS – TIMING OF APPLICATION

In winter cereals the relationship between shoot and ear development is less close than in the spring varieties. A range of post-emergence timings is possible in autumn and winter, depending on the herbicide(s) concerned, with a gap before spring spraying commences. In winter wheat ear initiation takes place in spring and the start of the safe stage for spring spraying, previously identified as the 'fully-tillered' stage, is now defined by the extension of the basal leaf-sheath – the 'pseudostem erect' stage (Fig. 5.2). Winter barley on the other hand initiates ears before the end of the year, especially if drilled early, so that the definition of satisfactory spraying guidelines is more difficult.

There is a lack of information regarding possible damage, for the newer chemicals especially, although there would appear to be a likelihood of leaf malformations resulting from early post-emergence use of MCPA and 2,4-D. Mecoprop on the other hand presents problems of leaf-scorch if frost occurs around the time of spraying, although there is usually ample opportunity for crops to outgrow this (Tottman, 1982).

The end of the safe stage for spraying in spring is again generally indicated by jointing, although some sources point to successful results at G.S.31 or 32 (first and second node visible, respectively), with mecoprop/contact mixtures especially. There appears to be general agreement, however, that spraying subsequent to G.S. 32 is liable to be harmful and that, in any case, delaying weed removal until these relatively late stages is almost certain to result in yield reductions, where competitive weed populations are present.

In summary therefore, there are essentially three stages when spraying may take place in winter cereals. These are:
(i) *Pre-emergence (autumn/early winter)*. This involves the use of residual herbicides applied to the soil in liquid or granular form after drilling. Specific restrictions apply to individual products in respect of timing, drilling depth, dosage rate and so on and, as is usual with such materials, soil type and condition are important both in relation to weed control and to crop safety. Herbicide performance is generally impaired by high levels of organic matter and also in some cases (e.g. isoproturon, chlortoluron) by the presence of surface trash and the residues of straw-burning.

The residual effect of the herbicides concerned varies, but in some cases (e.g. pendimethalin) is claimed to be sufficient to give control of spring-germinating species; a few chemicals (e.g. chlorsulfuron) also have an effect on emerged weeds.

(ii) *Early post-emergence (autumn/winter)*. These treatments are generally applied at around the two to three-leaf stage and, depending on the active constituents, may affect both emerged and germinating weeds. Both foliage-applied (e.g. mecoprop/cyanazine/HBN mixtures) and soil-applied (e.g. isoproturon/linuron/chlortoluron mixtures) are available. These treatments can be less expensive than the pre-emergence applications but difficulties may arise in relation to soil condition and the effects of frost.

(iii) *Late post-emergence (spring)*. This includes all treatments carried out between the pseudo-stem erect stage and G.S. 30 to 32, as previously discussed. The full range of contact/translocated mixtures is available, and the 'spraying window' is again usually about 10 to 14 days, although likely to be further constrained by rainfall and drift problems, with the growth-regulators especially.

There has been considerable debate over the years as to the relative merits of these timings. In terms of yield response, there seems to be little to choose between them, except where high weed numbers coincide with mild winters, when early removal of the weeds is likely to prove beneficial (Wilson, 1982). Levels of weed control, on the other hand, appear to be generally higher following autumn treatments in winter barley and spring treatments in winter wheat, which may necessitate a somewhat different approach to spraying in the two crops, where broad-leaved weeds are the major problem (Orson, 1982).

Further Reading

ELLIOTT, J. G. (1978), 'The economic objectives of weed control in cereals', *Proc. 1978 Br. Crop Prot. Conf. – Weeds*, pp.829-39.

Slightly dated in some respects, but still a valid and fresh overview of the topic.

MAKEPEACE, R. J. (1982, b), 'Broad-leaved weeds in cereals: Progress and Problems – a Review', *Proc. 1982 Br. Crop Prot. Conf. – Weeds*, pp. 493-502.

A broad-ranging review of various aspects of broad-leaved weeds in cereals, including distribution, effects on yield etc., and developments in control in winter crops especially.

ROBERTS, H. A. (ed.) (1982), 'Weed Control in Cereals', *Weed Control Handbook*, 7th edn., Blackwell Scientific Publications, Oxford, pp. 268-91.

Broad general account of the underlying principles. Includes coverage of weed effects, role of tillage systems, cereal growth stages, etc.; illustrated.

SNAYDON, R. W. (1982), 'Weeds and crop yield', *Proc. 1982 Br. Crop Prot. Conf. – Weeds*, pp. 729-40.

A thorough treatment of all aspects of weed/crop competition with particular reference to cereals. Stresses the need for planned approach to weed control and the long-term effects of varying management.

TOTTMAN, D. R. (1982), 'The effects of broad-leaved weed herbicides applied to cereals at different growth stages', *Aspects of Applied Biology 1*, 1982, Broad-leaved Weeds and their Control in Cereals, pp. 201-210.

Up-to-date review of the effects of growth-regulator herbicides in cereals. Useful introduction to this topic.

WILSON, B. J. (1982), 'The yield response of winter cereals to autumn or spring control of broad-leaved weeds', *Aspects of Applied Biology 1*, 1982, pp. 53-61.

Useful summary of the ongoing discussion re autumn/winter v. spring spraying.

6 Grass weeds – wild oats

6.1 Introduction

Historically, grasses have been less important than broad-leaved plants as weeds of arable crops. In recent years, however, certain species have become much more prominent, especially where continuous cereal-growing has been practised. Of these, couch or twitch is the most widespread and blackgrass the most competitive, while the greatest overall impact has probably been attained by the two species of wild oat. These established species are supplemented in different areas by invasive types such as barren and other brome-grasses, by pasture grasses such as perennial ryegrass and by volunteer cereals themselves.

A recent survey of the main arable areas in central and southern England highlights this regional variation, with the pasture grasses, predictably, more prominent in the west, couch in the east and wild oats in the south (Froud-Williams and Chancellor, 1982). All the major grass weeds, and especially wild oats, showed a decreased frequency compared with a previous survey in 1977, and winter barley was the cleanest cereal crop, with 40% of fields free of grass weeds as against 16% in the less competitive spring barley and winter wheat (Table 6.1). A wide range of herbicides is now available to control grass weeds, but their capacity for survival and the potential rate of increase demands the maximum attention to detail if control is to be successful.

A recent review of high-yielding cereal crops emphasised the durability of grass weeds, even in a relatively husbandry-conscious situation (Hollies, 1982). The performance of the major species from stubble to harvest was monitored over a five-year period. Meadow-grasses, couch and blackgrass were the most frequent species in the stubbles, while in growing crops wild oats was the commonest problem, followed again by blackgrass and the meadow-grasses. Most of the grasses concerned were reduced by good straw-burning but increased by high levels of nitrogen. Wild oats and the brome-grasses were commoner following reduced rather than conventional cultivations, while the reverse was true of the meadow-grasses and perennial ryegrass.

Herbicide efficiency was a major factor in determining yield. Over 50% of wild oat and over 40% of blackgrass treatments gave better

Table 6.1: *Grass weeds in cereal crops – % of cereal fields infested*

	Winter wheat	Winter barley	Spring barley
Wild oats	32	31	52
Rough meadow-grass	29	6	0.5
Couch-grass	26	20	53
Blackgrass	23	9	11
Barren brome	12	4	
Italian ryegrass	10	6	5
Bent grasses	9	5	4
Timothy	6	2	4
Onion-couch	4	1	2
Meadow brome	3	0.3	
Yorkshire fog	1	0.7	
Perennial ryegrass	1	0.3	
Soft brome	0.4	0.4	
Creeping soft-grass	0.5	0.1	
Loose silky-bent	0.3	0.1	
Marsh foxtail	0.2		
Smooth meadow-grass	0.1	0.1	
Wall barley	0.1		
Tall fescue	0.1		

(After Froud-Williams and Chancellor, 1982)

than 90% control, but 20% gave less than 80% control. This resulted in yield penalties in the worst cases equivalent to £65/ha for wild oats and £140/ha for blackgrass. There was a clear relationship between yield and weed prevalence at harvest, with an overall yield reduction of 17%, expressed mainly through a reduction in the number of ears/ m^2. This resulted in a gross margin difference between clean and very weedy crops of around 25%, taking both lost yield and control costs into account.

6.2 Wild Oats (*Avena spp.*)

6.2.1 IDENTIFICATION

The wild oats occurring in Britain include one tetraploid (*Avena strigosa*) and two hexaploids (*A. fatua* and *A. ludoviciana*). The latter, commonly known as 'spring' and 'winter' wild oats respectively, form part of an interbreeding series which also includes the cultivated oat, *A. sativa*. *A. strigosa*, the bristle oat, once cultivated in parts of western Britain, is now an occasional problem only in isolated parts of west-central Scotland.

Wild oats are readily distinguished from other cereal seedlings by the anti-clockwise rather than clockwise twisting of the leaves and by the absence of auricles – projecting teeth at the junction of the leaf-blade and sheath. Separation of wild from cultivated oats is more difficult at the seedling stage, although the former are likely to differ slightly in colour and are always hairy. As a general rule any cereal-type seedling appearing between the sown rows should be treated with suspicion. At heading the tall open panicles of wild oats are usually obvious in wheat or barley crops, while they may be distinguished from cultivated oats by the prominent black awns and the dense hairs at the base of the seeds.

Separation of the two main wild oat species depends largely on the method of dispersal of the seeds from the flower cluster or spikelet. Both species have two or occasionally three seeds per spikelet; in *A. fatua* all the seeds are awned, whereas in *A. ludoviciana* the third seed, where present, is awnless. At maturity in *A. fatua* the seeds are shed singly and each seed has a neat oval abscission scar at the base; in *A. ludoviciana* the spikelet is normally shed intact, only the basal seed possessing an abscission scar. In this case the remaining seeds, if separated for example by combining, show a small piece of the flower stalk or rachilla projecting from the base.

6.2.2. ORIGIN AND DISTRIBUTION

From centres of origin in Asia Minor the various wild oat species have now achieved a world-wide distribution. *A. fatua* predominates in temperate areas and is the main problem in north-west Europe and in North America. *A. ludoviciana* occurs mainly in areas with Mediterranean and similar climates.

As indicated earlier, *A. fatua* is a long established component of the British weed flora, and now occurs in most lowland areas. *A. ludoviciana* is of much more recent origin, first recorded in 1917, and is still largely confined to a relatively small area of central/southern England, associated with heavy soils and winter cereal growing. [A major source of information on all aspects of wild oat distribution, biology and control is *Wild Oats in World Agriculture* ed. D. Price-Jones (1981); see Further Reading.]

Wild oats only really emerged as a major weed problem following the increase in cereal-growing during the Second World War. Concern grew during the 1960s and early 1970s as evidence accumulated of the rapid spread of *A. fatua* into areas such as south Wales and north-west England, where it had previously been scarce or absent (Fig. 6.1). A survey of selected ADAS and other advisory regions of the UK in 1972 indicated that 55% of the area affected had become so during the

Heavy Infestation

Moderate Infestation

Fig. 6.1: *Wild oat levels and distribution in 1976.*

preceding six years and 85% during the preceding fifteen (Phillipson, 1974).

One result of this was the setting up of a national Wild Oat Advisory Committee, representing commercial and farming interests, research institutes and government departments. This body inaugurated a wide range of activities and publications, in an effort to increase farmers' awareness of the extent of the wild oat 'explosion' and of its effects on cereal production especially.

Despite this, another survey in 1977 revealed a further increase in wild oats, with two-thirds of all winter cereals affected and with areas such as eastern Scotland particularly hard-hit (Elliott *et al.*, 1979). More recent findings suggest that wild oats are now been contained in many areas, as a result, perhaps, of the effects of continued publicity, the availability of an improved range of herbicides and a better understanding of wild oat biology. Locally, however, spread continues, as in the south-west and in parts of Scotland, where long-standing pockets of heavy infestation also persist in, for example, Berwickshire, East Lothian and Moray and Nairn.

6.2.3 SEED SURVIVAL AND GERMINATION

The agro-ecology of wild oats centres on the role of the seed. The chief differences between the two main species, for example, relate to their differing germination patterns (Fig. 6.2). Spring wild oats germinate between March and early May, with a smaller flush from September to early November, while winter wild oats germinate throughout the October to March period, with a peak usually in November. Although most of the seeds emerge from the top 10cm. of the soil, they can germinate from greater depths – to over 20cm. in winter wild oats, with its slightly larger seed. In all cases it is the earliest-germinating seeds which are most competitive in cereal crops, producing the largest plants and, if not controlled, the maximum seed-return.

These germination regimes reflect a complex dormancy situation. In this country as many as 95% of *A. fatua* seeds are dormant at the time of shedding, as against a maximum of around 50% in *A. ludoviciana*, in which the first (lowest) seed in the spikelet is usually non-dormant and germinates within a few weeks of being shed. Dormant seeds of both species require a period of after-ripening or 'vernalisation' before they can germinate. The precise mechanism of this is unclear but several factors have been implicated, including low or fluctuating temperatures, soil moisture content and the presence of germination inhibitors in the seed itself.

It is difficult in practice to distinguish between this primary or 'natural' dormancy and the equally important secondary or 'enforced' dormancy which is responsible for the long-term survival of wild oat seeds. The latter is associated with a combination of factors inimical to germination, including low oxygen levels, as experienced by buried seeds. Burial may result from cultivation, frost action or, to a limited extent, from the self-burying action of the seed itself, brought about by the twisting response of the awn to alternate wetting and drying.

This reservoir of buried seed is important, in view of the ability of

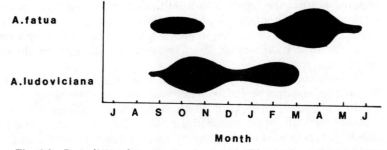

Fig. 6.2: *Periodicity of germination in spring wild oat (*Avena fatua*) and winter wild oat (*A. ludoviciana*) (after Potts, 1973).*

wild oats to build up, which makes prevention of seeding a priority. Selman (1970) found an average rate of increase in an uncontrolled population of 2.7/year, while Wilson and Cussans (1975) recorded a range of figures between 0.6 and 4.5, depending on level of control, straw treatment and cultivation regimes. Comparable rates of up to 2.45 in winter wheat and 1.3 in spring barley have recently been reported, reduced in the former to 0.83 by soil-applied herbicide and to 0.39 by late post-emergence treatment (Jarvis and Clapp, 1981).

However, the survival period of wild oat seeds in the soil is less than in many other weeds, at 8-10 years maximum for *A. fatua* and about half that for *A. ludoviciana*, under grassland or in undisturbed arable soil. Under normal arable conditions the rate of decline is even faster, with relatively few seeds surviving for more than three years, especially on lighter soils. Seedling production is greatest in the first two years after shedding, as the effects of natural dormancy wear off; older, less vigorous seeds and those emerging from depth are less likely to reach the surface and establish successfully. It appears, therefore, that in the south at least, the recurring problem of wild oats may be associated less with dormant seeds than with those freshly shed by plants not controlled in the crop (Wilson, 1981, a). Dormancy is evidently more prolonged in the cool, moist conditions of north-east Scotland, which may account in part for the extra vigour and persistence shown by wild oats in that area (Scragg and Kilgour, 1981).

6.2.4 GROWTH AND DEVELOPMENT

The initial development of the young wild oat seedling is slow and at this stage it does not easily withstand the effects of cultivation or crop competition. This may be due partly to its possessing only three seminal roots, as against four or five in wheat and at least six in barley. Between two and three weeks after germination, therefore, cereal seedlings have more extensive root systems than wild oats; later, the opposite is true, the crown or adventitious roots of wild oats being more vigorous. Shoot development follows a similar pattern, with tillering in wild oats being slow in the early stages, but increasing after about thirty days; on average about two or three tillers are produced per plant.

The wide spread of germination times in both species is reflected in an extended ripening period, with between three and five months from seedling to panicle emergence. In southern Britain panicles of winter wild oats normally appear around the end of June and seeds ripen in mid-July; spring wild oat is about two weeks later, especially in the north. Although cross-pollination can occur, wild oats are normally self-pollinated and isolated plants can therefore set seed – indeed initial infestations frequently appear to arise in this way.

The development of a schematic seed-cycle for wild oats (Fig. 6.3) highlights the importance of weed/crop interactions in determining eventual seed production. These factors, together with inherent variations in panicle size and tillering capacity, lead to marked fluctuations in the number of seeds produced, which can be anywhere in the range from 10-500 seeds per plant and up to 10,000 seeds/m^2. in suitable circumstances. In cereals a proportion of this seed is removed from circulation in the grain or straw, the amount varying with the season and the crop, but ranging, for example, from less than 2% in winter wheat and spring barley to over 17% in early-harvested winter barley (Wilson, 1981, a).

Post-harvest effects are also significant. Losses, especially in cereal stubble, can be considerable, partly due to the activities of birds, mice and soil organisms but largely to natural deterioration. Trials evidence over a number of years has demonstrated the desirability of leaving wild oat seed on the soil surface. The effects are compounded by the type of cultivation employed, the survival rate being lowest with direct-drilling and highest after deep ploughing, which places many seeds in a dormancy-promoting situation.

6.2.5 SEED DISSEMINATION

Much of the success of wild oats is due to their wide variety of farming-related dispersal mechanisms, the most significant of these being contamination of seed grain. The current Regulations governing seed purity are fairly stringent in this respect but when it is considered that even one seed/1000g sample can mean approximately 400 wild oat plants/ha the need for such control is obvious!

Work at the time of the wild oat explosion indicated the importance of farm-saved seed in this connection and results from the Official Seed Testing Station at Cambridge indicate that it is still a major problem. Over the three years 1978/79 to 1980/81 spring wild oat seed occurred in a mean of 12% of wheat and 16% of home-saved barley samples;

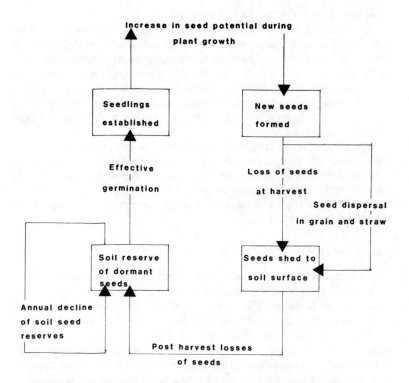

Fig. 6.3: *A schematic seed-cycle for wild oats. (after Wilson, 1981, a).*

this compared with averages of less than 0.3 and 0.6 for the corresponding crops in officially-derived samples, where the trend is currently downwards (Tonkin, 1982 – See Chapter 1, Further Reading, p. 25).

Another potent means of long-range spread is the contaminated straw bale, used for bedding and eventually returned to clean fields via the manure heap. A small, but significant, amount of seed can survive this treatment, especially if it is located in the drier and cooler parts of the heap. The same holds good for the small proportion of seed, about 5%, which survives digestion by livestock after being consumed in contaminated feed grain. In slurry, however, the survival rate is much lower, two to three weeks' immersion generally being sufficient to kill the seeds. Equally, wild oat seeds are unlikely to survive propionic acid or similar treatment. Finally, as regards long-distance dispersal, even grain sacks may carry a significant number of seeds, thanks to the 'adhesive' properties of the hairs and awns.

Local spread is associated particularly with farm machinery, both indirectly (for example in mud on wheels and tracks) and directly, in seed-drills and combine harvesters; the complex structure of the latter makes them especially liable. It has been a standard recommendation for many years that farmers and contractors should harvest clean fields first and that the combine and its associated trailers and balers should be thoroughly cleaned between jobs, preferably with industrial-type vacuum equipment.

The only 'natural' dispersal mechanism of any consequence is by birds, especially pigeons, which feed on cereal stubbles, but it appears that relatively few seeds survive their abrasive digestive processes. The lines of wild oat plants occasionally seen under power cables and the like are a result rather of beak-cleaning or preening, which dislodges externally transported seeds.

6.2.6 COMPETITION WITH THE CROP

The precise extent of this is often difficult to quantify under normal field conditions. Most investigations have been done in sprayed compared with unsprayed situations, which are complicated by the depressant effect of the herbicide and other factors. From a wide range of trials over many years, however, yield responses of up to 70% and more have been recorded, although 25-30% is perhaps more usual. For a more critical appraisal other methods of assessing yield reduction must be used, which may involve sowing seeds or planting seedlings of wild oats at a known density in crops, or else utilising naturally-occurring populations.

The greatest impact of competition, in spring barley at least, is clearly during shooting and earlier effects can be compensated for if the wild oats are removed before that stage. The situation in winter wheat is basically similar in that good yield responses to spraying can be obtained up until the second node is visible and before the flag-leaf has expanded, but not thereafter (Baldwin, 1979). It is widely recognised that early germinating wild oats can have an effect out of proportion to their numbers and that where these occur maximum yield benefits will be obtained by spraying as soon as possible (Peters and Wilson, 1983).

The factors affecting competition between wild oats and crops, reviewed by Chancellor and Peters (1976), can be summarised as follows:

(i) Date of sowing in relation to wild oat emergence – as previously discussed.

(ii) Different crops and cultivars. It appears that of the cereals barley is most competitive with wild oats, while no consistent differences appear among wheat, oats and rye. (Among other major field crops, beans appear to be particularly susceptible to competition, mainly on account of their open habit of growth).

(iii) Crop density and row width. There is wide agreement that increasing crop density is highly beneficial in suppressing wild oats, in terms both of shoot and seed production. Row width does not usually appear to have a significant effect.

(iv) Wild oat density. In general, the greater the wild oat density the greater the effect on the crop, although variable results have been reported. Long term studies by Jarvis and Clapp (1981) showed a significant relationship between wild oat numbers and yield reduction in winter wheat, but not in spring barley, which is more competitive.

Scragg (1980) emphasises, however, the difficulty in obtaining a clear response to different wild oat densities, due to the effects of variables such as date of emergence and crop vigour. So far as the latter is concerned, there is plentiful evidence of the beneficial effects of such factors as adequate fertiliser, pH and soil moisture levels in minimising competition. Other reports indicate a closer relationship between yield reduction and total wild oat dry weight at harvest (Wilson and Peters, 1982) or wild oat panicle dry weight (Baldwin, 1979), the latter suggesting an almost direct replacement of wild oat heads by cereal grains following a late post-emergence herbicide application.

Wild oats are typical of weeds which are troublesome before and after rather than during harvest. Being usually fairly dessicated by that time, wild oats do not markedly increase throughput and combine speed can be maintained without increasing grain loss. On the other hand, the presence of wild oat seed does increase the cost of the grain harvested (Table 6.2). The effect of wild oats in reducing the value of the grain itself is not in doubt. A contaminated seed crop is a serious liability and can lead to substantial losses through outright rejection or loss of premium, increased cleaning charges and loss of grain during cleaning, in addition to the reduction in yield (Elliott, 1978).

6.2.7 NON-CHEMICAL CONTROL

The control of wild oats may be attempted either directly, by hand-removal or by chemical and other means, or indirectly, for example by the use of clean seed or by crop rotation; in many cases a combination of these methods will be required. Three main situations can be envisaged: low wild oat numbers, typical of a developing (or declining) infestation; intermediate numbers, in some ways the most difficult to deal with; and high numbers, requiring urgent and perhaps repeated action. Recent reports suggest a reduction in frequency of the last, many infestations now falling within the intermediate category.

The distinction between technically and economically feasible weed control applies with particular force in the case of wild oats and it is highly desirable that measures are not undertaken haphazardly, but

Table 6.2: *Effect of wild oats on the cost of grain harvested*

Mean values from 51 experiments			
		Crop	
		clean	+ wild-oat
barley grain	t/ha	4.4	3.6
barley grain + straw	t/ha	8.2	7.0
wild oat grain + straw	t/ha	–	1.3
total combine material	t/ha	8.2	8.3
ratio barley grain: total bulk		0.54	0.43
combining cost	£/ha	49	49
combining cost, barley	£/t	11.1	13.6

(*After Elliott, 1978*)

rather as part of a definite and preferably whole-farm plan. Advisory guidelines formulated during the wild oat campaign identified a series of strategies under the broad headings of prevention, containment and eradication. Within this context, the methods available for the control of wild oats include the following:

(i) *Hand-roguing* This involves the uprooting of wild oat plants and their subsequent removal from the field for burning. Merely breaking off the heads induces the plant to produce tillers, while plants left lying in the field are very likely to set viable seed. Due to the extended ripening period of both wild oat species it is usually best to rogue twice, firstly when the heads appear above the crop in early to mid July and again some 10-14 days later, to catch late-maturing tillers.

Estimates of roguable populations vary, but are generally in the region of one wild oat plant to 20 or 30 m^2; small or patchy infestations can be rogued at higher densities. Roguing should be carried out in an orderly fashion, with teams of 4-6 preferably followed by a supervisor; for a variety of reasons it will probably be most efficiently carried out in the early morning or evening. The process can be greatly speeded up by the use of a chemical glove to apply herbicide, usually dalapon or glyphosate, to the wild oat seed head. This avoids the need to remove the plants, but the killed seed remains and may subsequently appear as an impurity in seed-crops.

(ii) *Straw-burning* Straw-burning works both by direct destruction of seeds (up to a maximum of about 30%) and also by breaking dormancy, so inducing seeds to germinate; the resultant seedlings can then be killed by herbicide or by cultivations. The effect is variable however, depending on how quickly burning is carried out after harvest, the amount of seed held in the straw and the temperature of the burn itself. Where seed from late-harvested crops especially has had time to penetrate the soil or simply to become covered by clods or stones, burning will have relatively little effect.

(iii) *Crop rotation and cultivations* It is appropriate to deal with these two aspects together, as it is largely the opportunity to vary cultivation techniques and timings which lends value to crop rotation as a weed control measure.

As an example we might take the situation where winter wild oats is the major problem. Given the relatively short dormancy of this species, good control may be achieved simply by a succession of one or more spring-sown crops, via appropriate stubble treatments or herbicides. The reverse situation is likely to be less effective for spring wild oats, however, especially with the current trend towards earlier sowing, which favours this species.

Elliott (1981) suggested alternating a winter cereal with spring barley and spring rape or any one of a selection of arable break crops (including arable silage, peas, beet, potatoes, maize or one-year leys). Of these, arable silage and short-term grass, cut before the wild oats have shed seed, are especially beneficial, while the remainder give scope for varied chemical treatment as well as seedbed and inter-row cultivations and hand-removal.

The effects of differing cultivation techniques on wild oat seed survival have been reviewed by Wilson (1981, a and b). Whatever the primary cultivation, relatively few viable seeds remain after three years, provided that further seeding is prevented. The rate of decline is faster, however, with tine cultivations than with ploughing. The beneficial effects of leaving freshly shed seed on the surface have already been noted, but the practicability of this is open to question. The same applies to the delayed sowing of spring cereals so as to allow maximum kill of emerged wild oats, which is almost certain to lead to unnacceptable yield reductions.

6.2.8 CHEMICAL CONTROL

At any level of infestation above the limit of hand-roguing, the use of chemicals has to be considered. For a number of years the choice of chemical in cereals was limited to one soil-applied type, i.e. tri-allate, and one foliage-applied, i.e. barban. These materials, while effective, pose certain difficulties of application and timing, but both are still widely used as relatively inexpensive alternatives to the newer generation of foliage-applied materials. The latter offer highly effective control of wild oats in all the main cereal crops, although some can display slight phytoxicity, expressed as mild scorching and stunting. Together with the older chemicals they provide a range of control options extending from pre- to late post-emergence of the crop.

The choice of chemical depends on the level of infestation and the desired result of containment or eradication. Late post-emergence spraying, by allowing maximum germination to take place, is more effective in controlling wild oats and is recommended where infestations are small enough to be unlikely to affect yield, or where eradication is the aim. Pre- or early post-emergence treatment, however, by removing the competitive early-germinating wild oats, may give a better yield response and is therefore indicated where infestations are high and/or a policy of containment is being followed. Success has been obtained recently with sequential treatments, involving both pre- and post-emergence chemicals, at normal or reduced dosage rates and the continued development of such techniques appears likely, especially where wild oats occur in conjunction with other grass weeds.

The stunting effect of some of these herbicides, leading to seed production below the crop canopy, can cause difficulties during subsequent hand-roguing, while in a few cases, such as barban and difenzoquat, certain cereal varieties are susceptible and therefore cannot be sprayed. Detailed information on these points and on herbicide choice, rates of application, timing and cost can be obtained from manufacturers or from M.A.F.F. Booklets 2252 and 2253, *Weed Control in Cereals*, updated annually.

6.2.9 ECONOMICS OF CONTROL

The difficulty of decision-making in wild oat control has already been noted. The crucial question is, at what level of infestation does chemical control become economically viable, bearing in mind the high cost of most wild oat herbicides? The answer must take into account not only the possibility of increased crop yield, but also considerations relating to harvesting, seed cleanliness and overall policy with regard to containment or eradication.

The overall effect of wild oats is not in question, but in practice, however, decisions relating to control have to be made on an individual unit basis and indeed may well vary from field to field. There are two thresholds involved: that at which yield becomes affected and that where an economic return over treatment costs may reasonably be expected. This has generally come to be regarded, in mainly cereal rotations, as around 12 wild oat plants per m^2; support for this figure has come, for example, from Jarvis and Clapp (1981), who found an economic threshold of 25-30 panicles/m^2, allowing for a mean of 2 to 3 panicles per plant.

Other experimental evidence indicates, however, that in many circumstances this figure may safely be exceeded. Haddow *et al.* (1978), for instance, suggested that for milling wheat or feeding barley a yield increase of .35 t/ha would offset spray costs, commensurate with a wild oat density of 60 panicles per m^2, approximately equivalent to 20-30 plants; higher prices for malting or seed crops would of course have justified spraying at lower levels.

Smith and Finch (1978) obtained largely negative margins in twelve out of sixteen trials with pre- and post-emergence herbicides against a range of infestations of 12-66 plants per m^2 in spring barley. On this basis they suggested that in this admittedly competitive crop, setting aside harvesting and cleaning costs, chemical control must be regarded largely as an insurance against continued increase in the wild oat population.

In winter crops, however, as infestations increase above the 12 plants/m^2 level, yields are increasingly likely to be under threat and chemical treatment therefore increasingly justified. It is in the 'grey area' below this level but above the limits of economic (and practicable) hand-roguing, that decision-making is most difficult and very much a matter of individual circumstances. At these levels yield responses, if any, are likely to be minimal, and the aim may well be one of containment, involving only such chemical inputs as will minimise seed-return to the soil and forestall population build-up.

Alternatively, if eradication is the objective, it will be necessary to continue chemical treatment until the level of infestation falls within roguable limits. It has been suggested (I.C.I., 1981) that a successful wild oat programme should last at least five years and achieve 95% control of wild oat seed return, while Large (1981) quotes one Cambridgeshire farm where a severe infestation took eight years of continued and varied chemical treatment to reduce to roguing level. It must be emphasised that the result of any such operation will be affected by herbicide performance (in limiting seed-production by survivors) and by cultivation technique; moderate control levels allied to tine or similar cultivations are unlikely to bring about a lasting improvement (Wilson, 1981, a).

Further Reading

HOLLIES, J. D. (1982), 'A survey of commercially grown, high yielding wheat and barley crops from 1977 to 1981', *Proc. 1982 Br. Crop. Prot. Conf. – Weeds*, pp. 609-18.

Novel approach, illustrating changes in grass weeds throughout the year.

JARVIS, R. H., and CLAPP, J. T. (1981), 'Effect of different herbicides and cropping sequences on the population dynamics of wild oats (*Avena* spp.) and on yields of wheat and barley', *Expl. Husb.*, *37*, pp. 133-43.

Useful bringing together of information, based on longer-term studies in E. Anglia but also of general relevance.

M.A.F.F. (1982), 'Wild -oats', *M.A.F.F. Leaflet No. 452*.

Concise account of wild oat identification, biology and control. Additional information re dosage rates, cost and application features of particular chemicals appears in M.A.F.F. Booklets 2252 and 2253, q.v.

PRICE-JONES, D. (ed.), '*Wild Oats in World Agriculture*', Agric. Res. Council, London, 1981, p. 296.
    An exhaustive review of wild oat biology, crop competition and control. Especially Chapters 1-5. Extensive bibliography.
WILSON, B. J. (1981, a), 'A review of the population dynamics of *Avena fatua* L. in cereals', *Proc. Grass Weeds in Cereals in the United Kingdom Conf.*, 1981, pp. 5-14.
    Up-to-date account of wild oat seed biology with reference to differing cultivation techniques and the role of herbicides.
WILSON, B. J. and PETERS, N. C. B. (1982), 'Some studies of competition between *Avena fatua* L. and spring barley: I. The influence of A. fatua on the yield of barley', *Weed Research*, 22, pp. 143-8.
    Representative of the more searching approach to weed/crop competition in cereals. Useful introduction to this topic. See also Peters and Wilson (1983).

7 Other annual and perennial grass weeds

7.1 Blackgrass (*Alopecurus myosuroides*)

7.1.1 IDENTIFICATION

An annual grass, also called rat-tail or slender foxtail, varying in height from about 10cm to 90cm. In competitive crops it may form only a single shoot, but will form tufts if allowed. The very characteristic flower spikes taper upwards and are often purplish in colour, hence its common name. Seedlings can be identified by the rounded tips to the leaves and by the remains of the typical, rather large (up to 7mm long), awned seed.

7.1.2 DISTRIBUTION

In recent years blackgrass has rivalled wild oats as number one arable weed in many areas. Following a decade or more of spread, it is now quite widely distributed, but with its centre of abundance still in south-east England (Fig. 7.1).

A survey carried out in 1977 (Elliot *et al.*, 1979) indicated that blackgrass was present on more than 50% of cereal-growing farms in England, comprising over 650,000 ha. Thanks to its mainly autumn germination it is closely associated with winter cereals, although spring crops can also be affected. Recent reports (e.g. Froud-Williams and Chancellor, 1982) indicate that blackgrass has to some extent been contained, although potential areas of expansion exist, for example in Scotland and in south-west England.

Moss (1980) has reviewed the agro-ecology and control of blackgrass and, unless otherwise indicated, references in succeeding sections are to his paper.

7.1.3 GERMINATION, GROWTH AND DEVELOPMENT

Blackgrass has a high potential for seed production, depending on the number of seeds per head (about 100, on average) and the number of heads per m^2 (up to 1000 in a very severe infestation). The seeds are

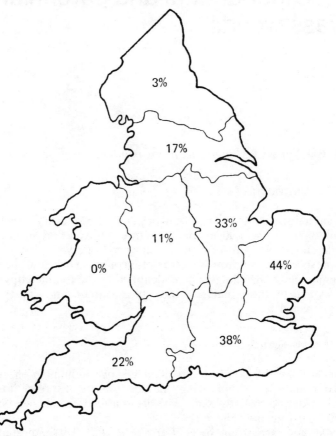

Fig. 7.1: *Percentage of the winter cereal acreage infested with blackgrass in A.D.A.S. regions in England and Wales, 1977. (after Elliott et al., 1979).*

shed over an eight-week period around late July, i.e. generally prior to the harvesting of winter wheat, but sometimes coinciding with that of winter barley, in which case many seeds are removed with the grain (Moss, 1983). Especially in the absence of cultivation, the death-rate of seeds reaching the soil-surface is considerable, due to attack by mice, birds and micro-organisms. Only one-third of viable seeds shed in late July survive until early October, most of this decrease taking place during the first two weeks after shedding (Fig. 7.2).

On shedding, blackgrass seeds show a short-lived natural dormancy, so that most (> 80% in ideal conditions) will germinate in the same autumn. However, particularly if they are buried by cultivation or if the soil is waterlogged or very dry, enforced dormancy may prevent germination until suitable conditions recur. Evidence is conflicting as to the length of time blackgrass seed will withstand burial at various

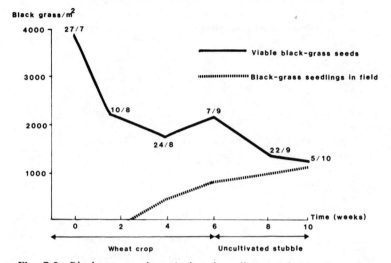

Fig. 7.2: *Blackgrass seed-survival and seedling production in cereal stubble. (after Moss, 1980).*

soil depths, estimates ranging from a few months to several years. Close to the soil surface many seeds can survive at least until the following spring and if the main autumn germination is retarded the normal small spring flush can expand to become a serious threat. The plant is relatively tolerant of high levels of soil moisture, hence its association with heavy land or lighter soils with impeded drainage or a high water-table.

By far the greater number of seedlings emerge from the top 2.5–3cm of the soil. Autumn-germinating seedlings grow slowly during the winter, and in average UK conditions most survive, those which do not succumbing as much to frost-heave as to low temperatures. In spring, growth picks up very rapidly and the weed is then well placed to enter into competition with the cereal or other crop. Fig. 7.3 summarises the annual seed-cycle.

7.1.4 COMPETITION WITH THE CROP

There is no doubt that blackgrass offers severe competition to cereal crops and marked yield reductions have been reported on many occasions. Most of the competition is due to blackgrass seedlings emerging after drilling, supplemented on occasion by earlier-germinating plants which may be in effect 'transplanted' at drilling, if not killed off previously. It would appear that the main period of competition follows the commencement of active blackgrass growth in the spring, and it is essential that the weed is removed before this stage, if meaningful yield responses are to be obtained.

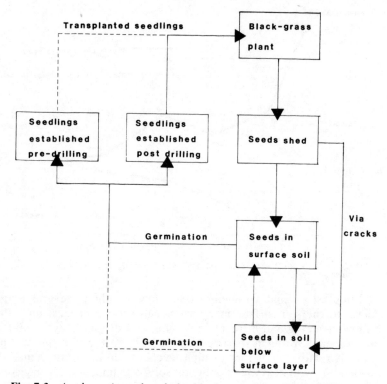

Fig. 7.3: *A schematic seed-cycle for blackgrass. (after Moss, 1980).*

Competition appears to be mainly for nitrogen and results specifically in fewer ears and a smaller number of grains per ear. The effect of a given blackgrass infestation in terms of reduced grain yield is difficult to predict and is closely related to the vigour of the crop; this, if adequate, can effectively suppress even a heavy infestation, whereas a weak or thin crop will allow the weed to express its tillering potential to the full – as many as 150 tillers per plant have been recorded in such situations. Blackgrass is much less competitive in spring cereals and yield reductions in these are less likely to be significant.

As with wild oats, much of this yield loss assessment has involved responses to herbicide treatment. Baldwin (1981) reviewed several years of ADAS investigations into this and other aspects, and concluded that 'no very satisfactory relationship was established between blackgrass infestation and likely yield response to herbicide use'. Nonetheless, from this and related studies in the UK and elsewhere, it appears that yield reductions are likely if there are > 30 blackgrass plants/m^2 in early spring, while the figure of 50–55 plants/m^2 has been fairly widely accepted as a base level above which a positive return from normal herbicide application can be expected.

7.1.5 NON-CHEMICAL CONTROL

The development and eventual containment of a blackgrass infestation involves quite complex interactions between a variety of husbandry factors and aspects of the biology of the weed. The basic importance of crop vigour in combating blackgrass has been stressed, and Moss lists four points affecting this which may be manipulated to the advantage of the crop:

(i) *Seed-rate.* Although experimental evidence is scarce, it appears likely from first principles that higher seed-rates will be beneficial.

(ii) *Fertiliser.* The addition of nitrogen generally favours the crop, although blackgrass does respond, and where waterlogging or other limiting factors occur may do so preferentially.

(iii) *Crop variety.* Here again it would seem reasonable to assume that high-tillering varieties are more able to withstand competition.

(iv) *Drainage.* Blackgrass tolerates waterlogging better than do cereals and it appears that its spread during the 1960s and 1970s may be in some degree attributable to the succession of relatively mild, wet winters experienced during that period.

That apart, however, it is clear that a major factor in the recent blackgrass upsurge has been the widespread adoption of minimal cultivations in cereal crops. Such techniques favour blackgrass by leaving freshly shed seed on or near the soil surface, precisely the area where, as previously indicated, almost all the weed's germination occurs. Fig. 7.4 illustrates the effects of various cultivation methods on the distribution of blackgrass seed in the soil in autumn, together with an indication of the number of seedlings subsequently emerging in the crop.

The harmful effects of these techniques are compounded by the trend towards earlier drilling, which means that many more seedlings develop within the crop, protected from cultivations and herbicides and also by the fact that current herbicide practice still does not prevent the return of a substantial quantity of blackgrass seed to the soil. On the other hand, the immediate benefits of ploughing may be at least partly offset in the longer term by its tendency to bring dormant seed to the surface; if seed return could be prevented, minimal cultivation methods might ultimately be more advantageous.

Straw-burning has also been shown to have a marked effect on blackgrass numbers. Depending on the thoroughness of the burn, up to 70% of seeds may be destroyed, while others are induced to

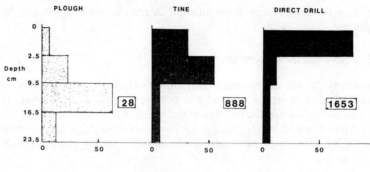

% Seed in depth fractions of soil (Nov)

Fig. 7.4: *Effect of cultivation system on the incorporation of blackgrass seeds in the soil. Numbers of blackgrass plants emerging subsequently are shown in the boxes. (after Moss, 1979).*

germinate, rendering them open to attack by other methods. In a long-term experiment at WRO, Moss (1981) investigated the effects of cultivation and straw-disposal treatments on blackgrass build-up. In general, ploughing resulted in only a slow increase, irrespective of whether the straw was burned or baled-off. With tine-cultivation or direct-drilling, however, burning greatly reduced build-up, and would appear to be an essential adjunct to these techniques.

7.1.6 CHEMICAL CONTROL

Chemicals offering a measure of blackgrass control first became available in the late 1960s, when changing husbandry patterns were leading to increased interest in the grass weeds. Herbicides specifically directed against blackgrass soon followed, and by 1977 almost 40% of the cereal acreage infested was being treated (Elliott *et al.*, 1979). The properties and performance of the materials concerned have been widely reported over the years in BCPC Symposium Proceedings and elsewhere, while the results of numerous ADAS trials up to 1980 have been reviewed in detail by Baldwin (1981). The main features of those chemicals currently approved for blackgrass control are given in MAFF Leaflet No. 522.

Of the established herbicides, chlortoluron and isoproturon have fairly consistently given the best control of blackgrass, although this is not always fully expressed in crop yield, due to an element of phytotoxicity in these materials, which are also relatively expensive. With chlortoluron especially, early post-emergence application appears to give the best results, while isoproturon, a somewhat more soluble type, is preferred for early spring use and where chlortoluron-sensitive cultivars are involved.

On balance it appears that the performance of these primarily soil-active materials is reduced where minimal cultivations or direct-drilling are regularly employed and especially where straw residues or ash are left on the surface; the presence of burned straw residues has been shown to increase markedly the adsorptive capacity of the upper soil layers (Cussans *et al.*, 1982). In these circumstances the contact/residual herbicide diclofop-methyl has given good results.

Among other candidate chemicals, metoxuron alone or with simazine has produced results comparable to those obtained with isoproturon, while methabenzthiazuron, although giving poorer overall control, shows almost as high a yield response, reflecting perhaps its lower phytotoxicity. In newer chemicals, pendimethalin gives very good control pre-crop emergence and similar results have been obtained pre- and post-emergence with chlorsulfuron, in mixture with methabenzthiazuron (Upstone *et al.*, 1982). Certain wild oat herbicides also show useful control of blackgrass in suitable circumstances, but these are unlikely to be preferred, especially where it is intended to minimise seed return to the soil.

The twin constraints of difficult soil conditions and limited time often experienced on the more intensive cereal units have led to the use of reduced-volume applications of isoproturon and other chemicals. Generally favourable results have been obtained with these techniques, whose potential may be increased with the wider introduction of low ground pressure vehicles (4.2.2). Another new approach involves the sequential application of pre and/or post-emergence herbicides at reduced dose, with or without the addition of surfactants. Although results have been variable, the technique can be more cost-effective than full-rate applications in any one year and may offer possibilities of reduced crop damage and generally improved flexibility (Ayres, 1982).

Overall, the results of field experience and trials alike reflect the complex nature of the crop response to the use of herbicides against blackgrass, involving such factors as weed numbers, time of emergence of crop and weed, crop vigour and the presence of other grass or broad-leaved weeds. Most herbicides will give 'commercial' control of around 75% to 85%, however, and the more effective types 95% or better.

It is clear that, as with wild oats, decisions relating to the control of blackgrass should be made with definite objectives in view. Non-significant levels of infestation can be achieved and maintained with regular use of the more effective herbicides, but at a cost which may be excessive in many seasons. (In this connection, Rule (1981) points out that control of a visually unacceptable infestation may not provide a worthwhile yield response). Alternatively, a variety of low-cost strategies are becoming available, which will often give quite satis-

Table 7.1: *Control strategies in blackgrass*

Control factor	Best strategy	Worst strategy
Cultivation	Plough	Direct drill or reduced cultivations
Herbicide	Good herbicide at correct timing	No herbicide
Straw disposal	Spread and burn	Bale
Drilling date	Late	Early
Stubble hygiene	Complete weed kill	Incomplete weed kill
Rotation	Include non-cereal crops	Continuous winter cereals
Crop competition	High seed rate; High rate of N fertiliser; Vigorous cereal variety; Drainage programme	Low seed rate; Low rate of N fertiliser; Weak growing cereal variety; No drainage programme

(*After Moss, 1980*)

factory results, although the regenerative capacity of blackgrass must always be kept in mind; the various options are summarised in Table 7.1.

7.2 Other annual grasses

7.2.1 BROME-GRASSES (*BROMUS* SPECIES)

Few other annual grasses seriously threaten arable crops. The most prominent, apart from volunteer cereals themselves, are the brome grasses, mainly barren or sterile brome (*Bromus sterilis*) and to a lesser extent meadow brome (*B. commutatus*) and soft brome (*B. mollis*). These are essentially plants of headlands and hedgerows, which have recently emerged rather rapidly as weeds of winter cereals, especially in Lincolnshire, the Midlands and the Cotswolds, with a frequency in these areas of about half that of blackgrass (Froud-Williams and Chancellor, 1982).

The reasons for this increase are not altogether clear, but may include close harvesting of field-edges, hedgerow removal, the use of flail-cutters and especially the practice of cultivating headlands to a shallow depth as a fire-break prior to straw-burning. Whatever the reason, having gained entry these weeds are clearly favoured by certain aspects of cereal monoculture as currently practised and where numbers are high are very competitive, at least on a par with blackgrass. An intensive research effort sparked off by this situation has thrown light on the seed biology and general behaviour of these species (e.g. Budd, 1981).

Sterile brome is an autumn-germinating winter annual. The seed has almost no natural dormancy and mostly germinates soon after it is shed. Germination is inhibited, however, by drought and by low temperature after shedding and it appears to be a combination of these factors which produces the well-marked spring flush frequently shown by this weed, in which, unusually, seed-burial does not result in dormancy (Froud-Williams, 1981). Indeed, given that the seed is unable to emerge successfully from a depth of more than 10cm in the soil, it can easily be seen how suited the plant is to minimal cultivation systems.

There appears to be general agreement that no one herbicide is capable of giving consistent control of sterile brome in cereals, at least with a single spray. To get round this, sequences have been examined with the intention of catching both the autumn and spring flushes, the most successful perhaps being those involving tri-allate followed by one or other of the blackgrass herbicides, notably metoxuron. This sort of programme is clearly not cost-effective in most cases, and must be seen as a means of forestalling a potentially damaging build-up.

Rule (1981) describes a number of chemical and other alternatives and emphasises in particular the value of deep cultivations or mould-board ploughing, at least on headlands, as a means of burying seed below germination depth. Another possibility is to delay autumn drilling long enough to allow maximum germination, the resultant seedlings being readily killed during seed-bed preparation. Finally, the presence of this weed is another argument for the inclusion of a suitable break-crop (for example field beans or oilseed rape), in which herbicides such as TCA and propyzamide offer more efficient and less costly control than is currently obtainable in cereals.

7.2.2 CANARY-GRASS (*PHALARIS PARADOXA*)

The latest 'problem' species to come to prominence is *Phalaris paradoxa*, an introduced plant related to the common coastal and wetland species reed canary grass (*P. arundinacea*). First noted in any quantity in 1978 in coastal areas of Essex and Suffolk, it has now extended its range more or less due westwards across the Midlands and into S. Wales. Concern has been expressed that the species will follow the well-worn path involving reduced cultivation systems and the use of farm-saved seed, and such experimental and observational work as has been carried out appears to support this. Although the biology of *P. paradoxa* has not been explored in any detail, it appears to germinate mainly in autumn and is therefore well adapted to winter-sown crops.

Canary-grass is somewhat resistant to isoproturon and chlortoluron, but is susceptible to autumn applications of non-urea herbicides such as pendimethalin alone and terbutryne or diclofop-methyl alone or following tri-allate (Martindale and Livinstone, 1982). Ploughing is also evidently a very useful control measure. It appears that if farmers and advisers are alive to this problem it ought to be possible to prevent its becoming yet another headache for the cereal grower.

7.3 Integration of annual grass-weed control in cereals

The control of these annual grass weeds is a major issue in cereals especially, with nearly 2,000,000 ha (almost half the cereal area) sprayed in 1981. Despite this the problem has not gone away; wild oats may have been suppressed to a certain extent but blackgrass is at least holding its own and sterile brome continues to extend its range. It will be clear from the preceding sections that three factors stand out in this situation – more winter cereals, more reduced cultivations and more early drilling; Cussans (1981) points out that current cereal-growing techniques are so favourable to these weeds that anything short of full control is likely in many cases to do little more than maintain the status quo.

The chemicals now available offer the possibility of controlling these weeds over a wide range of crop growth-stages, so that given suitable spraying conditions, treatment is possible in every month from September until May. However, taking all relevant factors into account, some basic guidelines can be established. With blackgrass, for example, the main urea-type herbicides are clearly most effective in autumn and spraying with these should take place preferably before New Year and at the latest by the end of January in wheat and rather earlier in barley. In wild oats on the other hand, the position is less clear. Pre- or early post-emergence treatments may be less expensive and offer yield advantages by early removal of competition, but spring-treatments can yield as well (especially in wheat), give better weed control, and fit in better with other commitments.

Where both weeds are present the main objective must be to attack the blackgrass first, in the autumn, picking up the remaining wild oats with a spring application (most blackgrass herbicides give some control of wild oats). Sequences may be based for example on tri-allate and isoproturon followed by one or more applications of full or half-rate difenzoquat, while combinations with metoxuron may be preferable if sterile brome is present. Alternatively, promising results have been obtained with single applications of tri-allate and isoproturon in granular formulation (Atkin and Turner, 1982) and such mixtures are attracting increasing attention.

A central issue here is the question of eradication as against containment and the supposed 'threshold-levels' for the major species have been discussed. Infestations substantially exceeding these levels clearly must be controlled to prevent immediate and heavy yield losses, but at or below the threshold decision-making becomes more difficult and must also take into account harvest and post-harvest effects and the possibility of seed-return, leading to increased problems at a later stage.

Reference has already been made to the case for a co-ordinated approach to grass weed control, involving the use of chemical and cultural methods, together with rotations. The use of break-crops such as oil-seed rape or winter beans offers substantial benefits in terms of varied herbicide usage and cultivations, including ploughing, while peas, arable silage and spring barley additionally provide an opportunity to control autumn-germinating species. Spring barley in particular enables relatively inexpensive control of grass weeds and has lower variable costs all round, a major consideration in present circumstances.

7.4 Perennial grass weeds

7.4.1 IDENTIFICATION AND DISTRIBUTION

Perennial grasses have long been established as weeds of arable land. They typically possess well-developed underground creeping stems or rhizomes and have often been referred to as couch-type grasses, after common couch or quickens (*Agropyron repens*), historically the most widespread and troublesome species. Table 7.2, from the review by Boyall, Ingram and Kyndt (1981), summarises the main features of the species concerned. In addition to those listed, onion-couch or false oat-grass (*Arrhenatherum elatius*), a non-creeping type with persistent annual bulbs at the base of the stem, can also cause problems occasionally.

These grasses are widely distributed throughout Britain, although common couch predominates in East Anglia, northern England and Scotland, where it has tended to be favoured by intensive cultivation and spring sown crops. It frequently occurs in mixed populations with black bent (*Agrostis gigantea*) and is to some extent replaced by it on thin, chalk soils. Creeping bent (*A. stolonifera*) occurs on a range of soil types but is usually associated with impeded drainage. It is a more prostrate species with creeping leafy stolons and to that extent is somewhat more amenable to control by herbicides.

7.4.2 GROWTH AND DEVELOPMENT

The biology and control of these grasses centres round the creeping stems, which provide both a means of spread and an overwintering foodstore. It is important to remember that even the underground types are *stems*, and therefore possess buds along their length. Control has always been compounded by the difficulty of getting at these

Table 7.2: *Summary of the main features of the couch-type grasses*

Name	Reproductive parts and overwintering state	Depth	Reproduction by seed
Agropyron repens (common couch, twitch, wrack, wicks)	Rhizomes with dormant underground buds, aerial shoots overwinter	Shallow 7cm.	Moderately important
Agrostis gigantea (black bent, twitch, wrack)	Rhizomes with dormant underground buds, overwinter and can root at nodes	Shallow 5cm.	Very important
Agrostis stolonifera (creeping bent, fiorin, white bent, watergrass)	Aerial creeping stems (stolons) that overwinter	Above ground	Importance unknown
Holcus mollis (creeping soft grass, soft fog)	Rhizomes with dormant underground buds. Aerial shoots overwinter.	Mostly 5cm. deep but can be deeper	Not known
Poa trivialis (rough-stalked meadow-grass)	Creeping stolons. Foliage overwinters in a reduced form.	Above ground	Important

(*After Boyall et al., 1981*)

protected structures.

Most investigative work has been done on common couch. Its growth season extends from March to October, when most aerial shoots die back, although a few may be produced during mild winters. Rhizome production proceeds evenly throughout the spring and early summer, decreasing in late summer and autumn as the aerial shoots – the food source – die back. Cropping, especially with strong competitors such as rape or spring barley, can markedly delay rhizome production, placing greater emphasis on late-season development.

The rhizomes are mainly confined to the top 15cm of the soil, forming a more or less dense mat. Ploughing to a depth greater than 15cm is therefore an integral part of the more traditional control systems, while reduced cultivations or direct-drilling favour couch and are generally to be avoided.

The role of the rhizome buds is of great importance. In undisturbed conditions about 95% will be dormant at any given time, held so by the action of hormones produced at the growing point of the rhizome. This 'apical dominance' is released if the rhizome is broken by cultivations and manipulation of this operation, although to a large extent overtaken by recent herbicide developments, remains a potentially valid means of controlling couch in appropriate circumstances.

Seed production occurs from mid July onwards, after the harvest of many cereal crops in the south at least. Freshly shed seed can germinate immediately, but in fact does not usually do so until conditions are suitable, which may not be until the following spring. The extent to which couch-type infestations develop from seed was probably underestimated in the past, but is now recognised as being of some importance, especially in black bent. Rough-stalked meadow grass has the peculiarity that only vernalised seedlings produce viable seedheads, so that spring-sown crops offer an opportunity to reduce this species.

7.4.3 COMPETITION WITH THE CROP

There are somewhat conflicting reports as to the competitive effects of these grasses. Common couch, which produces rather sparse foliage, is probably the least aggressive; quite high densities have to be reached before cereal yields are significantly affected and certainly before any economic return may be expected from control. Cussans (1970), for example, found that 45 shoots/m^2 did not significantly affect cereal yield, while 180 shoots/m^2 produced a 20% reduction, the effect being greater in spring than in winter crops. Equally, Scragg (1980) showed that the competitive effect of couch is small compared with that of wild oats, and that with fewer than 100 shoots/m^2 only small effects on yield were recorded.

Despite such findings and the variable results obtained in control experiments over the years, it is clear that the higher infestations do pose a threat to cereal yields. This is especially true of those species, such as rough-stalked meadow-grass, which produce a relatively dense mat of foliage, which if nothing else is likely to cause significant problems at harvest, in both cereals and root crops.

With common couch in particular, another consideration is that of disease carry-over; Table 7.3 lists those cases which have been substantiated. In addition to this, more direct effects on crops have also resulted from the action of toxins released by the decaying rhizomes, in common couch and creeping soft-grass especially. This feature, known as allelopathy, has been demonstrated both in pot trials and in field situations, in a variety of crops including cereals,

Table 7.3: *Pests and diseases associated with perennial grass weeds*

Species	Pest
A. repens	Stem and bulb eelworm (*Ditylenchus dispaci*)
A. repens	Cereal root eelworm (*Heterodera major*)
A. stolonifera	Ear cockles (*Anguina agrostis*)
Various, particularly	Frit fly
A. repens	Wheat bulb fly (*Heptohylengia waritata*)
	Gout fly(*Chlorops purilioris*)
	Hessian fly
	Wireworms (*Agriotes* spp)
	Cereal aphids, various spp.
	Takeall (*Gaeumannomyces graminis*)
	Ergot (*Calviceps purpurea*)
	Powdery mildew (*Erysiphe graminis*)
	Barley leaf blotch (*Rhyncosporium secalis*)
	Rusts (*Puccinia* spp)
	Smuts (*Ustilago* spp)

(*After Boyall et al., 1981*)

lucerne and maize. The effect is a complex one, involving specific decomposition products of the rhizomes and competition from micro-organisms involved in the breakdown process.

7.4.4 CULTURAL/CHEMICAL CONTROL

The control of these mainly rhizomatous grasses has always been something of a problem. In pre-herbicide days the solution was often to bury the rhizomes below subsequent cultivation depth, or alternatively to try to bring them to the surface and kill them by desiccation or by repeated cultivation of emerged shoots. On heavy land especially, these methods were often used in conjunction with full or partial fallowing. The technique persists in the form of stubble-treatments using cultivations and/or herbicide to kill off regrowth. This system requires the maximum possible growth-period for best results, so that early harvest followed by efficient straw removal or burning are essential prerequisites.

Initial cultivations are carried out with a rotavator or similar implement in order to obtain maximum rhizome fragmentation and hence shoot formation from the rhizome buds. Further passes are then made at about 2–3 week intervals or when the couch shoots have reached approximately 5cm (two-leaf stage). Treatment not later than this stage has the effect of gradually exhausting the food reserves stored in the rhizome. At least three cultivations are likely to be required, followed by deep ploughing to bury the weakened rhizomes.

It should be noted that unless conditions allow the full programme, all that will be achieved will be a further spread of viable rhizome fragments throughout the field.

These initial cultivations have also proved necessary in couch-control systems involving the use of herbicides, which became available from about the late 1950s onwards. The simplest expression of this involves the use of paraquat, alone or with diquat, as a replacement for cultivations to kill couch regrowth. This again must be carried out by about the two-leaf stage of the weed for maximum benefit.

Stubble-cultivation can also be a prelude, however, to the use of translocated herbicides, i.e. aminotriazole and (in England and Wales only) dalapon, which are applied to the foliage of actively-growing couch grass some 4–6 weeks after harvest. Spring applications are also possible, but are more likely to interfere with following crops and are in any case somewhat less effective. Aminotriazole is more effective against common couch (and some broad-leaved weeds) and dalapon against black bent. Specific time intervals must elapse between spraying and the sowing of susceptible crops.

As an alternative to these, TCA may be incorporated into the soil in spring or autumn, again following shallow ploughing or cultivations or as a one-pass system involving the use of a combined sprayer and rotavator; there are well defined safety intervals prior to the sowing of following crops. Brassicas are least susceptible and at lower dosage rates may be sown 5–7 days after application; potatoes, peas and beans require up to 8 weeks and cereals 12–16 weeks, which effectively means a spring cereal following an autumn treatment. Bulb crops are particularly sensitive to TCA and should not be sown or planted until a full year has elapsed.

7.4.5 CONTROL WITH GLYPHOSATE

The arrival of glyphosate in the late 1970s added a completely new dimension to couch-control in cereals. The initial use of this actively translocated, non-persistent material was as a foliage-application in cereal stubble any time from early autumn onwards. In this case no post-harvest cultivations are called for, to allow for maximum regrowth, which must be more extensive than with the older chemicals, at 10–15cm minimum height (five expanded leaves). The typical yellowing or reddening of the foliage appears from 10 to 50 days after spraying, depending on the season; cultivations leading to sowing the next crop can commence immediately afterwards.

More recently a novel application of glyphosate has been developed whereby the chemical is applied to couch foliage in wheat or barley just before harvest, when the crop is fully ripe but the weed is still green and actively growing. Treatment can take place as soon as the moisture content of the grain falls below 30%, and at least seven days must then be allowed before harvest begins. This procedure has given excellent control of couch and a range of other weeds, including later-developing perennials such as thistles and field bindweed (O'Keefe, 1981). There are no restrictions on feeding the grain or straw following this treatment and the technique has recently been cleared also for specialist uses such as malting and milling, but not for the seed trade.

A recent appraisal of the technique in Scotland (Sheppard et al., 1982) largely confirmed initial favourable impressions. Couch control of between 95% and 99% was achieved in the first year, compared with a maximum of 75% with stubble treatments and the yield increases more than covered the cost of the treatment, including wheeling effects. Some yield penalties ensued when severe conditions of wind and rain occurred between application and harvest, and there was one case of reduced germination in grain treated at the higher rate. However, benefits were also obtained in terms of lower grain losses or, alternatively, higher throughput, and an added feature was the consistent reduction in grain moisture content.

Interest has grown recently in the use of reduced-rate versions of this approach as a means of improving conditions for straw-burning. In general it appears that for maximum reliability of control full-rate applications are desirable, but that half-rate treatments will give acceptable and indeed equivalent results in many situations. However, the effects of still lower doses, adequate perhaps for 'cleaning-up' purposes, are very dependent on ideal conditions for herbicide uptake and translocation. The addition of surfactants (not at the time of writing approved by ACAS) may compensate in some cases for a reduced dosage-rate, but a more important prerequisite for successful control at all rates of application appears to be the presence of sufficient moisture in the soil to ensure adequate weed growth for the herbicide to be effective (Orson, 1982).

7.4.6 MEADOW-GRASSES (POA SPECIES) – BIOLOGY AND CONTROL

Two of these grasses are extremely common in cereals and other crops throughout most of the UK. These are rough-stalked meadow-grass (P. trivialis), a perennial, and annual meadow-grass (P. annua), a short-lived 'opportunist' species prevalent in stubbles, but capable of growing and flowering during most of the year.

In central/southern England rough-stalked meadow-grass was found to infest an area similar to that of the main grass weeds in winter wheat (29% of fields examined) but was much less abundant in winter and especially spring barley (Froud-Williams and Chancellor, 1982). Its pattern in this survey was most like that of blackgrass or sterile brome (Table 6.1). Its distribution is more even, however, in the north and in

Scotland, where it is co-dominant with couch in the main arable areas and is probably increasing.

The meadow-grasses have received relatively little attention compared with some other species, but they can be quite competitive, associated with yield reductions of .3t/ha in barley and .5t/ha in wheat if not controlled (Hollies, 1982 – see Chapter 6, Further Reading, p. 111). Other evidence suggests that this can probably be attributed mainly to *P. trivialis*; annual meadow-grass, although often visually prominent after harvest, tends to develop late in the life of the crop and in general poses little threat to yield.

Ryegrasses and, more locally, timothy have also been noted recently as an increasing but still relatively minor problem. Italian ryegrass in particular averaged around 8% occurrence in winter crops and 5% in spring barley in Froud-Williams and Chancellor's cereal survey, and the level of this and the other species concerned is probably significantly higher in areas of more mixed farming. Hollies found very similar levels of ryegrass in high-yielding winter wheat crops, associated with yield reductions of between 5% and 13%.

Ryegrasses and other pasture grasses are readily controlled by a variety of herbicides, including glyphosate, TCA or tri-allate pre-sowing and barban and diclofop-methyl post-emergence. In most cases best control will be obtained at the seedling stage.

Further Reading

BOYALL, L. A., INGRAM, G. H. and KYNDT, C. F. A. (1981), 'A literature review of the biology and ecology of the rhizomatous and stoloniferous grass weeds in the U.K.', *Proc. Grass Weeds in Cereals in the United Kingdom Conf.*, 1981, pp. 65–76.

Broad-ranging review, covering identification, distribution, dormancy and growth, including the effects of environmental factors, crop yield effects and toxin production.

CUSSANS, G. W., MOSS, S. R., HANCE, R. J., EMBLING, S. J., CAVERLY, D. J., MARKS, T. G. and PALMER, J. J. (1982), 'The effect of tillage method and soil factors on the performance of chlortoluron and isoproturon', *Proc. 1982 Br. Crop Prot. Conf. – Weeds*, pp. 153–60.

Initial stage of an in-depth examination of the factors underlying reduced activity of these two very important herbicides.

MOSS, S. R. (1980), 'The agro-ecology and control of blackgrass (*Alopecurus myosuroides* Huds.) in modern cereal-growing systems', *ADAS Quarterly Review, 34*, pp. 170–91.

Detailed discussion of all aspects of blackgrass biology and control in relation to current cereal-growing practices.

M.A.F.F. (1981 et seq.), 'Blackgrass', *M.A.F.F. Leaflet 522*.

Covers all aspects of the biology and control of this major grass weed. Should be read in conjunction with Moss (1980), above.

O'KEEFE, M. G. (1981), 'The control of perennial grasses by pre-harvest applications of glyphosate', *Proc. Grass Weeds in Cereals in the United Kingdom Conf.*, 1981, pp. 137–44. Useful introduction to this increasingly important technique.

8 Weed control in row crops

8.1 Introduction

For two hundred years after they were introduced in the eighteenth century these crops formed the basis for weed control in the farming rotation. The concept of growing crops in rows, so permitting the control of weeds by inter-row cultivations, served agriculture well until the middle of this century. However, the continuing run-down and increasing cost of the agricultural labour force after 1945 placed increasingly severe constraints on these crops, which, without adequate weed control, are extremely susceptible to competition.

Further pressure has been exerted through the development of mechanical harvesting in a number of crops and by the need in others (e.g. vining peas, oilseed rape) for rigorous standards of cleanliness in the marketed product. Without the introduction of suitable herbicides it is doubtful whether some of the more mechanised crops would have been able to reach their present position. The main difficulty lies in devising herbicides capable of selecting between crop and weeds. This is especially true where the two are botanically related, as with cruciferous weeds (charlock, runch, shepherd's purse) in brassicas, fat hen in sugar beet and nightshades in potato.

Because of this the herbicides available for these crops have comprised mainly pre-sowing or pre-emergence residual types, which place less strain on selectivity. This does involve a risk of crop damage in lighter soils if drilling depth is incorrect or heavy rain follows application, and, conversely, the risk of herbicide failure in dry seasons. These materials also show generally reduced effectiveness in organic soils, due to binding of the herbicide to the soil colloids. Over the years the position has improved with the gradual introduction of foliage-applied, post-emergence chemicals, especially in sugar beet and oilseed rape, while the most significant recent development has been the introduction of materials such as alloxydim-sodium and fluazifop-butyl, which offer enhanced grass-weed control in the growing crop.

Novel methods of handling the available herbicides have also been developed in recent years, often initiated by farmers and growers themselves in an attempt to circumvent problems of restricted weed spectrum and cost. These have included 'band-spraying' of the crop rows in sugar beet and some brassicas, the sequential use of pre and/or post-emergence chemicals (often at reduced application rates), and the use of a wide range of approved and 'unofficial' tank-mixtures. Another new technique involves the use of rope-wick and similar applicators, which utilise translocated herbicides to selectively control tall-growing weeds.

One particular problem which may be mentioned here is that of volunteer crop plants, acting as weeds in other crops. This can involve broad-leaved crops in cereals, cereals in root crops and herbage grasses in both! The problem appears to have increased in recent years and such volunteers can occasionally form a part of the weed spectrum, particularly in intensive cereal and cereal/grass rotations or with potatoes (Cussans, 1978). Generally speaking, although some of the more intractable cases may have been eased by means of the new graminicides for example, the intensification of cropping in many arable areas means that the problem is likely to continue.

8.2 Brassica crops

8.2.1 FORAGE BRASSICAS

These have always been the 'poor relations' among crops so far as chemical weed control is concerned, and have suffered more than most from post-war pressures of labour and costs, allied to the difficulty of devising suitable selective herbicides. Some of the crops concerned, such as kale and forage rape, are quite competitive and especially if grown in summer when few weeds are germinating, may encounter relatively few problems. Spring-sown turnips and swedes on the other hand are slow to establish and cover the ground, and yields may be severely affected if weeds are not controlled. The weeds involved are mainly annuals such as chickweed, charlock, runch, fat hen, mayweeds and the polygonums. Couch-type grasses are a common problem, as are species such as docks and creeping thistle, where the brassica has been direct-drilled into a grass sward.

A restricted range of soil-applied pre-sowing and pre-emergence residual herbicides is available for the control of annual weeds, supplemented by a few post-emergence treatments. The activity of the former is strongly affected by soil moisture levels subsequent to application, while the pre-sowing types, in addition, have to be incorporated into the soil, to avoid losses from volatilisation or photochemical breakdown. All of the chemicals require attention to detail in terms of dose and timing of application and all have shortcomings of weed spectrum, which has given rise to a limited use of mixtures and sequential applications. With swedes, the existence of a domestic (i.e. higher value) market and the possibility of mechanical harvesting have meant renewed interest in chemical weed control and promising results have been obtained with mixtures and sequences of new and existing herbicides (Cromack and Davies, 1982).

Herbicide treatments in these crops are therefore as follows:

(i) *Pre-sowing treatments.* The main material involved here, trifluralin, represented something of a breakthrough when it was introduced in 1967 and, despite problems of limited range, has been a major factor in brassica-growing during the intervening period. Timing (within thirty minutes) and thoroughness of incorporation is fairly critical and this must be to such a depth that subsequent cultivations will not bring untreated soil to the surface. This is especially important where crops are grown on ridges, in which case the incorporation depth must be doubled.

Trifluralin gives good control of many annual weeds but not composites (e.g. mayweeds, corn marigold), cruciferous weeds or cleavers. The related chemical dinitramine offers a somewhat wider control spectrum and also a longer period after spraying during which incorporation must take place; neither material is recommended for use on sandy or organic soils. A mixture of trifluralin and napropamide (Neepex*) is also available for use in turnips and swedes. The combination controls a wide range of weeds, but is more expensive than the basic chemicals and its extra persistence, while beneficial in an average season, has given rise to residue problems under cold, dry conditions when breakdown is inhibited.

(ii) *Pre-emergence residual.* Here again one herbicide predominates, i.e. propachlor, introduced at about the same time as trifluralin and available as a liquid or granular formulation. This material is easier to apply and more tolerant of high soil organic matter (i.e. above 8–10%) than the pre-sowing types, which are, however, much less expensive. Its weed spectrum is complementary in that it gives good control of trifluralin-resistant species but not of the polygonums, fumitory, fat hen or orache. Sequential application with trifluralin is possible and has given good results in trials with swedes. The cost of the propachlor treatment may be reduced by band-spraying of granules, especially in ridge-grown crops. Propachlor can also be used post-emergence, but only if the weed seedlings themselves have not emerged and not before the three or four leaf stage of the crop.

Similar constraints apply to the tank-mix of propachlor and chlorthal-dimethyl, which controls more weeds but is again considerably more expensive than the basic material. Two new chemicals, tebutam and metazochlor, have also shown promising results in swedes and may provide useful alternatives to the existing treatments.

(iii) *Post-emergence.* Treatments of this type have largely been confined to kale, the waxy leaves of which offer improved protection against herbicide penetration. Desmetryne and to a lesser extent sodium monochloroacetate are the main chemicals used, either singly or in sequence with trifluralin. Composite weeds and charlock are resistant to the combination, the latter having proved intractable to all conventional brassica herbicides and mixtures (Cromack and Davies, 1982). The only other recommended post-emergence treatment, aziprotryne in cabbages, does control this weed, but not cleavers or runch.

Another approach to weed control in brassicas, offering perhaps the best chance of controlling resistant weeds, is the so-called 'stale-seedbed' technique. This has been used mainly on light sandy soils or peats, unsuitable for normal pre-emergence treatments. Basically the seedbed is prepared in the normal way and then left until a flush of weed seedlings appears; these are then sprayed off with paraquat, sowing taking place some two or three days afterwards. Later emerging weeds are not affected and may require further mechanical or chemical control.

As previously indicated, perennial grass weeds can be a serious problem in brassica crops. Control is best carried out at the pre-sowing stage, with glyphosate or TCA. Glyphosate is considerably more expensive, but also controls broad-leaved weeds; it will most likely be applied pre-harvest in a preceding cereal crop. TCA may be incorporated into the soil up to within seven days of sowing and offers adequate control of couch and other such grasses on most soil types.

Annual grasses, including volunteer cereals, can also be a problem. Control of these in future may rest with post-emergence applications of the new graminicides such as alloxydim-sodium. These also offer some control of perennial grasses, although at the higher rates required there is some evidence of risk to the crop, at least in swedes (Lawson and Wiseman 1982).

8.2.2 OILSEED-RAPE

The expansion of this crop shows no sign of levelling off, having increased from around 5,000 ha in 1970/72 to around 175,000 ha in 1982, an area almost equal to that of potatoes or sugar beet (Jones and Orson, 1982). The weed problems are those which might be expected in an autumn-sown crop, including annual meadow-grass, blackgrass and couch-type grasses, chickweed, mayweeds and speedwells. Cleavers can be a problem at harvest-time and as a seed-impurity, as can charlock and runch, the presence of the seeds of which can lead to rejection of the crop. Another, often major, problem concerns volunteer cereals, especially following winter barley.

Given the overall extent and value of the crop, it is not surprising to find that oilseed rape supports a much wider range of herbicides than any of the other brassicas. Pre-drilling incorporated treatments include trifluralin on its own or with napropamide against broad-leaved weeds, tri-allate against wild oats and TCA against grass weeds generally. TCA is also widely used at the pre-emergence stage against annual grasses and volunteer cereals, but has recently come under fire on the grounds that it reduces the amount of surface wax on the rape leaves, predisposing the crop to subsequent herbicide damage or to attack by fungal diseases such as *Leptosphaeria* and *Alternaria* (e.g. Gladders and Musa, 1982). It is, however, much less expensive than the alternatives, which are also somewhat more demanding in terms of timing and soil condition.

Pre-emergence control of broad-leaved weeds has centred in the past mainly on propachlor, supplemented recently by metazochlor and tebutam. Propachlor has only a short residual life in the soil and is therefore perhaps more suitable for crops in the north and in Scotland, where drilling is later and weed germination stops earlier in the autumn. Metazochlor on its own has given excellent control of the major dicot. and grass weeds, except volunteer cereals; it should not be applied when the oilseed rape is germinating. Metazochlor is generally, and tebutam only, recommended to be used in tank-mixtures or, preferably, in sequence with pre-drilling TCA, for improved control of wild oats, sterile brome and volunteer cereals.

A wide range of post-emergence treatments is available, based mainly on propyzamide and carbetamide, singly or in mixtures with, for example, benazolin and clopyralid, the choice of chemical depending largely on the main weeds present. Clopyralid has specific activity against composite weeds, including mayweeds and corn marigold, but neither it nor its mixture with benazolin controls annual grasses. All these herbicides have their own properties in terms of weed spectrum, crop and weed growth-stage and, where appropriate, soil and weather criteria, which must be taken into account for best results. Full details of these and of dosage rates, costs, etc. may be obtained from MAFF Booklet No. 2068, 'Weed Control in Oilseed Rape'.

Adequate levels of weed control are therefore clearly attainable in this crop; the extent to which this is reflected in increased yield is, however, less obvious. Ward (1982) obtained significant yield responses in only four out of eleven trials involving a wide range of pre- and post-emergence treatments. Because of the varied weed flora and the relatively long life of the crop, a convention has developed of sequential pre- and post-emergence applications, often based on TCA; however, Ward again found no significant benefits from sequences as against single herbicides with a comparable weed spectrum. Finally, a major consideration in choosing a herbicide in this crop relates to herbicide residues. The more persistent materials such as trifluralin and propyzamide should be avoided where there is any real likelihood of crop failure, for whatever reason.

Jones and Orson (1982) have suggested that oilseed rape will perhaps be the main outlet for the new graminicides, given that about 80% of the crop is currently sprayed with TCA, and to a lesser extent dalapon, against annual grasses and volunteer cereals. In theory this mainly pre-emergence, 'insurance' spraying might be replaced by autumn post-emergence application of a graminicide, with a resultant gain in flexibility. Fluazifop-butyl especially has given improved control and also significant yield responses, especially in conjunction with a propyzamide application. Whether these increases are sufficient to offset any increased cost is again unclear, and may depend on whether the graminicide can be applied early enough to pre-empt grass weed competition (see also 8.6, below).

8.3 Sugar Beet

8.3.1 WEED PROBLEMS

In sugar beet especially, radical changes in weed-control practice took place over a relatively short space of time. The crop was conventionally grown with a great deal of hand-labour for singling and weeding, backed up by a variety of mechanical gappers and thinners, with interrow cultivations to control weeds between the crop rows. The declining and increasingly costly labour force of the 1950s and 1960s, together with advances in other aspects of sugar beet husbandry, placed a premium on the development of suitable selective herbicides. Following the introduction of pre-emergence chemicals in the early 1960s and the post-emergence contact material phenmediphan some years later, there was, therefore, a very rapid increase in the area treated from about 10% in 1961 to over 80% by 1969 (Fig. 8.1). A wide range of single chemicals and mixtures is now available, offering control of most weed permutations.

Among the broad-leaved species fat hen, chickweed and the polygonums are especially prominent, while there is quite a marked divergence between the weed flora of mineral soils and that of the fen peats on which much of the crop is grown, the latter supporting much higher populations of a smaller range of species. Annual grasses are often a problem, especially where beet is grown in rotation with cereals, and a recent survey indicates that more than half the sugar beet acreage has significant amounts of wild oats and blackgrass. Couch-type grasses can also be a nuisance, particularly in fodder beet and mangolds, which tend to have a more northern and western distribution. As in other root crops, weeds can severely affect yield and also interfere significantly with harvesting and subsequent handling of the crop.

A major weed problem over the last few years has been 'weed beet', comprising a variety of annual forms which, like bolters, flower and set seed in the growing crop and cause severe harvesting and other

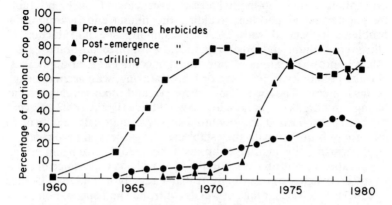

Fig. 8.1: *The progress of herbicide usage in sugar beet, 1960–1980. (after Roberts, 1982; courtesy of the British Sugar Corporation).*

problems. Weed-beet infestation has increased steadily from around 14% of sugar beet fields in 1977 to around 27% in 1982. The effect is cumulative in the sense that about 30% of seed shed survives in the soil each year. Buried seed can last for at least 7 and up to 15 years; its persistence is therefore favoured by deeper cultivations, which place it below germination depth.

It is essential therefore that weed beet is controlled before viable seed is shed, which is usually by mid-August or the end of July in earlier seasons. Ninety-five per cent control or better can be obtained by three cuts at suitable intervals during the flowering season, while the main hope of chemical control may rest with the application of glyphosate in rope-wick and similar applicators, which take advantage of the height differential between the weed and the crop.

8.3.2 WEED CONTROL

For many years chemical weed control in sugar beet and related crops depended on a limited range of herbicides, including lenacil, chloridazon and propham/chlorpropham/fenuron mixtures (soil--applied, mainly pre-emergence or pre-drilling) and phenmedipham (contact post-emergence). Later, trifluralin was introduced for pre-drilling or post-emergence use and in 1977 metamitron became available over a wide range of crop growth stages. Clopyralid and ethofumesate give improved control of certain broad-leaved and, in the latter case, also grass weeds, including sterile brome and volunteer cereals.

Wild oats can be controlled (and blackgrass suppressed) by tri-allate pre-drilling or diclofop-methyl post-emergence, and couch preferably by pre-harvest glyphosate, avoiding the timing and incorporation problems associated with TCA or dalapon. The introduction of alloxydim-sodium and other graminicides, applied post-emergence, offers improved flexibility, wider-ranging control, easier application and the possibility of localised or band-spraying, with accompanying savings in cost. They will also control rye and other cereals used as 'living wind-breaks' in areas liable to wind-blow (Breay, 1982). Recent work has emphasised the potential of these materials, alone or in mixture or sequence with the established broad-leaved herbicides.

Experimentation in order to broaden the weed spectrum and extend the duration of control has led to the approval of a wide range of tank-mixtures, which are increasingly available readily-formulated. Details of cost, dosage-rates, range of weeds controlled and application times etc. are contained in MAFF Booklets 2254 (*Weed Control in Sugar Beet*) and 2256 (*Weed Control in Fodder Roots etc.*) and useful summaries are published periodically in the farming press and in the *British Sugar Beet Review* (e.g. Bray, 1983). It should be noted that the soil-applied chemicals in particular are subject to the usual limitations of soil-type, rainfall etc., while residual effects also have to be watched, especially in dry seasons or following crop failure.

For various reasons, not least the relatively high cost of many of the chemicals involved, there has always been considerable interest in more or less specialised application techniques in sugar beet. One expression of this has been the widespread use of 'band-spraying' – spraying only the crop rows with herbicide, the remaining weeds being controlled by inter-row cultivations. Alternatively the weeds between the crop rows may be treated with a directed paraquat spray, using shields to protect the crop plants, a slow and somewhat risky technique. The value was soon recognised of sequential applications of pre- and post-emergence herbicides, in enhancing both the effectiveness and the duration of weed control and by the late 1970s such programmes were established as standard practice at least on mineral soils.

On the peats, on the other hand, which because of their adsorptive properties are basically unsuited to the use of pre-emergence herbicides, chemical weed control has largely been restricted to post-emergence treatments. Recently, however, developments have been initiated in these areas which have to a degree revolutionised the approach to chemical control in sugar beet. These involve the use of repeated low doses of phenmedipham or metamitron, applied post-emergence, with the object of catching successive weed germinations as they occur. The chemicals are applied overall in a reduced water volume, which in itself provides substantial operating benefits.

Initially, high spraying pressures were seen as being fundamental to the success of this 'low-dose/low volume' (LD/LV) technique, in enabling the production of abundant fine droplets and hence better coverage. Latterly, however, comparable results have been obtained at normal pressures, provided that suitable spray-nozzles are used and even where the total amount of herbicide is less than would have been applied in a single full-rate application. Spraying at the susceptible cotyledon stage has given excellent results on both peat and mineral soils, although in the latter case circumstances such as early weed germination may still require the use of a pre-emergence herbicide. Weed control, crop safety and sugar yield are all marginally better with metamitron, partly, perhaps, because phenmedipham tends to produce crop scorch in hot, sunny weather.

This essentially farmer-led development is now being increasingly supported by manufacturers' recommendations and a variety of avenues are currently being explored with a view to exploiting the full benefits of the new approach.

8.4 Potatoes

8.4.1 WEED PROBLEMS

The development of weed control in the potato crop has been broadly similar to that in sugar beet. The choice of herbicide is more limited, but the chemicals involved are effective and clean crops are feasible with a quite high degree of reliability. The crop is of course grown on ridges, and weed control has traditionally involved much hand-work, supplemented by tractor-drawn implements, which control weeds either in the process of forming the ridges or subsequently.

The repeated passage of cultivation machinery through the crop has a number of harmful side-effects, however, resulting mainly from the erosion of soil cover from the tubers. These include:

(i) physical damage to the tubers
(ii) increased risk of frost damage and partial greening of exposed tubers
(iii) increased moisture loss from disturbed soil
(iv) likelihood of disease transmission from leaves to tubers
(v) increased clod formation, especially on heavier soils, leading to problems with mechanical harvesters

Early work in East Anglia (Mundy, 1975) showed a fairly substantial yield benefit in favour of herbicides, with improved moisture retention being perhaps the most telling factor.

For these and for more general reasons there was a rapid increase in the use of herbicides when suitable materials became available in the 1960s, particularly in the early potato areas, where growers were evidently quick to see the advantages. In 1953 only about 2% of the crop was treated, but this rose to 44% in 1960 and to nearly 70% in

1973. Since then the position has stabilised somewhat, since some growers continue to use mechanical methods and also some potatoes are grown on soils unsuited to residual herbicides. In general the costs of herbicides and cultivations for weed control more or less balance out and where suitable machinery has been available farmers have tended to use it.

8.4.2 WEED CONTROL

Most herbicide applications in potatoes are pre- or early post-emergence residual. Their effect is therefore mainly on germinating weed seedlings, but many also have some contact effect on emerged weeds. Selectivity depends on creating a herbicidal barrier in the upper layers of the soil, well above the rooting zone of the potato plants. Where these herbicides have been used, therefore, further cultivations should be avoided, in order that the herbicide seal remains intact and untreated soil is not exposed on the sides of the drills. The residual effect is mostly fairly limited, so that there are advantages in delaying application until the latest recommended stage, provided only that emerged weeds fall within the contact range of the herbicide concerned. At the same time residue problems can arise in dry summers, especially in early potatoes, which have of course a much shorter growing season.

The chemicals concerned are subject to the usual restrictions in terms of soil-type/dosage rate. In highly organic soils their effect is reduced, while in very light soils there is a risk of crop damage, particularly in high rainfall areas. On the other hand, unless there is some rain following spraying, weed control may be reduced, as it will be if contact between the herbicide and the weeds is prevented by a coarse or cloddy tilth.

The candidate herbicides include linuron, chlorbromuron and prometryne with a mainly residual action, and dinoseb and cyanazine with a predominantly foliage effect. Dinoseb is a scheduled (Part II) substance under the Agriculture (Poisonous Substances) Regulations (Appendix 2), and suitable precautions should be observed when using it. Mixtures of linuron with trietazine and terbutryne with terbuthylazine are also available. Metribuzin, a translocated/residual herbicide of improved selectivity, can be applied pre-emergence in early potatoes and pre- or post-emergence in main crops (except Maris Piper and Pentland Ivory), until the potato shoots are 15cm. (6") high. It is also recommended as a pre-planting incorporated treatment in fen peat soils.

The main alternative to these treatments, both for broad-leaved and grass weeds, lies in a pre-emergence application of paraquat, on its own or with diquat, to kill emerged weed seedlings and above-ground shoots of couch and other perennial grasses. Results depend on maximum weed germination prior to spraying, hence it is probably more suited to main-crop than to early potatoes. Application may be delayed until up to 10% emergence in early potatoes and 40% in main-crops, with a view to enhancing weed kill. The emerged potato shoots, which must not be more than 15cm, will be scorched, but the plants usually grow away from this setback quite quickly. Fluoroc-hloridone, the first new potato herbicide for some time, is currently awaiting full PSPS clearance.

Glyphosate and TCA are also available to control annual and perennial grasses pre-planting and dalapon similarly pre-emergence. The specialist potato herbicide EPTC will control couch in the fortnight prior to planting, while diclofop-methyl is effective post-emergence against wild oats. The only other post-emergence treatment, MCPA against late-germinating annuals and some perennial broad-leaved weeds, must be regarded as an emergency application, and crop damage is likely especially in seed crops, where the resultant leaf-symptoms may result in disqualification.

More recently the graminicides have become available for post-emergence grass weed control. Alloxydim-sodium is recommended against a variety of annual grasses, including volunteer cereals in ware potatoes only; it also provides useful suppression of couch at around the three or four leaf stage. Fluazifop-butyl has been found to give significant yield increases following couch control, particularly where preceded by an application of paraquat to take out emerged annual weeds. There is of course no residual effect, so that there may be a need for cultivation between the drills to catch late-germinating weeds.

8.4.3 HAULM DESICCATION

Another main use for herbicides in the potato crop is to stop haulm growth with a view to hardening off tubers before harvest, which is also facilitated by the desiccation of weed growth; the treatment also reduces the spread of blight (*Phytophthora infestans*) spores to the tubers. Four main chemicals are involved; of these sulphuric acid is the quickest-acting and is recommended for seed production, while dinoseb and especially diquat are liable to cause internal browning and heel-end rot in tubers if applied in very dry conditions. Metoxuron is the slowest-acting of the four, but has the advantage that certain fungicides may be mixed with it.

8.5 Crop Legumes

8.5.1 WEED PROBLEMS

The crops include broad and field beans (*Vicia faba*), dwarf or French beans (*Phaseolus vulgaris*), runner beans (*P. coccineus*) and dried and vining peas. Although these are fairly uncompetitive, at least in the early stages of growth, the harmful effects of weeds are due as much to harvesting and marketing problems as to yield reductions, especially at lower infestations.

Problem weeds include smothering types such as chickweed, knotgrass and speedwells; tall-growing species, including fat hen, redshank, charlock and corn marigold; and climbing species such as cleavers and black bindweed. Field bindweed, sow-thistle, creeping thistle and other perennials are competitive during growth and a nuisance at harvest, especially in processed crops, as are volunteer potatoes and black nightshade. Fruits and seeds of these and other species which are difficult to separate from the harvested product can lead to crops being rejected.

8.5.2 WEED CONTROL

Current practice in these crops generally requires a fairly high level of herbicide usage, involving mainly pre-emergence residuals. There is some use of post-emergence treatments, especially in peas, against the later-developing perennial weeds, but the best control of these, where appropriate, is probably pre-harvest glyphosate. This also applies to perennial grasses, although alloxydim-sodium and especially fluazifop-butyl now offer at least suppression of these in the growing crop.

Control of annual grasses is available via broad-spectrum chemicals such as carbetamex and propyzamide, while for wild oats the choice is between the inexpensive but somewhat demanding tri-allate and post-emergence diclofop-methyl. Here too, however, the graminicides, with their wider range and flexibility of timing, appear to offer significant advantages.

8.5.3 BROAD BEANS AND FIELD BEANS

These are large-seeded crops, sown relatively deeply, and represent a good example of herbicide selectivity by 'depth protection' – the control of germinating weeds by a layer of herbicide separated from the crop seeds by untreated soil. Their open growth makes them susceptible to early weed competition, although later they do develop some smothering effect. This, together with the fact that these crops are rather intolerant of post-emergence herbicides, means that pre-emergence treatments are generally preferred.

Simazine has for many years been the standard herbicide against annual broad-leaved weeds, on its own or in mixture with trietazine. Terbutryne and chlorpropham mixtures are also available, while propyzamide additionally gives some control of volunteer cereals and the major annual grasses, in winter crops only. Simazine can also be applied post-emergence, again in winter beans only, as can car-betamex, which also controls annual grasses, and dinoseb-acetate. Various formulations of trifluralin may also be used, depending on the crop concerned; generally speaking these require to be followed by a post-emergence treatment for best results.

8.5.4 DWARF BEANS AND RUNNER BEANS

These crops are usually grown in narrow rows, making mechanical control difficult, and are at the same time relatively uncompetitive, so that herbicides are an essential feature of their husbandry. A range of pre-emergence chemicals is available, including a specific material, diphenamid, which can also be used post-emergence in a tank-mixture with chlorthal-dimethyl. There is some potential in rope-wick treat-ments to take out tall-growing weeds such as fat hen.

8.5.5 PEAS

This is another crop which as currently grown is very dependent on the use of herbicides, so much so that particulars of these are often included in growing contracts. This relates to the effects of weeds in hindering the mobile harvesters and interfering with the drying of the crop. In peas, the presence of those weeds whose seeds or fruits are difficult to separate from or cause taints in the produce are a particular nuisance. Weeds can also affect yield, sometimes severely, and yield improvements from the use of herbicides are commonly found (Lawson, 1983).

Being very early sown the crop plants usually emerge before the weeds, so that contact pre-emergence treatments are unsuitable. At this time, too, soil conditions may make the value of residuals uncertain, although useful chemicals are available, including pro-metryne, cyanazine and terbutryne and trietazine mixtures on normal soils and aziprotryne and chlorthal-dimethyl plus methazole on sandy soils.

If pre-emergence control is not feasible or less than fully effective, a range of post-emergence treatments is possible, including most of those used in field beans, plus cyanazine or bentazone with MCPB/MCPA and dinoseb acetate and amine. MCPB alone is preferred against thistles, docks and field bindweed, although these and other perennials are probably best controlled in or following the preceding crop.

The same is true of perennial grasses although, as in beans, suppression can be obtained with alloxydim-sodium and especially fluazifop-butyl. With couch and wild oats, significant yield responses have been obtained only when weed numbers were high or the crop relatively non-competitive. Annual grass-weed control is as in beans, but with peas, which are often grown on rented land with unpredictable grass-weed problems, the post-emergence graminicides may prove particularly advantageous. Peas appear to be generally tolerant of these chemicals at normal application rates and they have also been successfully integrated into sequences with pre- and post-emergence herbicides for broad-leaved weed control, although tank-mixtures have proved less successful.

The question of tolerance is an important one in the pea crop, where the susceptibility of cultivars has been found to vary in relation to a wide range of herbicides, depending on the waxiness of the leaves. Anything which reduces the amount of wax, such as frost, wind-damage, leaf-diseases or the effect of pre-sowing TCA, can place even resistant varieties at risk (King, 1980). A simple test is available which enables growers to test the amount of leaf wax in cases where doubt exists. The Processors and Growers Research Organisation (PGRO, Thornhaugh, Peterborough PE8 BHJ) has produced Information Sheets covering 'Pea Leaf Wax Assessment', 'The Reaction of Pea Varieties to Herbicides' and 'The Choice of Herbicides for Peas' and herbicide labels also carry the requisite information. A similar leaflet, from the same source, covers the reaction of broad and dwarf bean varieties to herbicides.

8.6 Grass weeds in row crops

The occurrence of grass weeds in broad-leaved crops is difficult to quantify although there are signs of increase in winter cereal areas, especially in autumn-sown crops such as oilseed rape and winter beans, where volunteer cereals themselves often constitute a major part of the problem. In the past the tendency has been for chemical control of perennial grasses especially to be carried out before planting or drilling; the main herbicides involved, including glyphosate, amino-triazole, dalapon and TCA, have been discussed in a previous section (7.4.4). Prior to autumn-sowing, the preferred treatment is now probably pre-harvest glyphosate in the preceding cereal, although the full rate is more expensive than other chemicals.

The other materials mentioned all require back-up cultivations, including reasonably deep ploughing, for best results, followed by a minimum safe interval before sowing or planting susceptible crops. This may be anything from a few days (e.g. TCA before swedes) to six weeks or more (e.g. full-rate dalapon before sugar-beet). They can all be used prior to autumn or spring crops, but the safe interval necessary and the length of time required to carry out the full treatments militate against their use in northern areas.

Table 8.1: *Advantages and disadvantages of pre-as against post-emergence treatments for the control of grass weeds in row crops.*

Advantages of pre-emergence or pre-drilling herbicides

---

1. Weeds are controlled before they emerge and before they can compete with the crop.
2. Residual herbicides are usually of a sufficient persistence in the soil to control a prolonged germination of a grass or a range of grass species.

Disadvantages of pre-emergence or pre-drilling herbicides

---

3. Incorporation may delay drilling, seedbeds may be over-cultivated in wet conditions and the extra tractor wheelings may create soil structure problems.
4. There may be a time interval specified between application of the herbicide and drilling of the crop.
5. If there is a crop failure, subsequent choice of cropping may be limited or delayed.
6. Seed may have to be drilled at a specified depth.
7. Dry soil may limit herbicidal action.
8. Organic matter absorbs residual herbicides, limiting their use to soils below a specified level.

Advantages of post-emergence herbicides

---

9. Weed populations can be assessed before spraying.
10. Spot spraying or band spraying may be employed under certain circumstances.

Disadvantages of post-emergence herbicides

---

11. Crop may shade target weeds.
12. Timing in relation to the use of other crop protection inputs may be a problem in some circumstances.

(After Jones and Orson, 1982)

Limited control of annual grasses is available from selective herbicides such as simazine in beans, ethofumesate in sugar beet and propyzamide in oil-seed rape, but the number of non-specific materials is limited. Tri-allate is the most widely used, mainly against wild oats; it is relatively inexpensive and the use of a granular formulation can offset problems of incorporation. Of the post-emergence materials barban, benzoylprop-ethyl and diclofop-methyl also give good control of wild oats, while the last is moderately effective against a range of other species.

There has long been a need, therefore, for the development of efficient selective post-emergence herbicides against grass weeds in the row crops, and it appears that this is now being met by the introduction of the graminicides. At the time of writing only alloxydim-sodium (Clout*) and fluazifop-butyl (Fusilade*) have full commercial clearance in the UK, but the range and effectiveness of these chemicals is likely to increase as new analogues become available.

Table 8.1 summarises the advantages and disadvantages of post-emergence as against pre-emergence and pre-drilling herbicides for the control of grass weeds in row crops and suggests that, on balance, the former may be more suitable for annuals and the latter for perennials. The matter hinges to a large extent on the timing of control relative to the onset of weed competition, which is likely to be greatest in autumn and early spring-sown crops. For annuals especially, the graminicides appear to offer significant improvements in control, although where infestations are particularly high a pre-emergence chemical may still be necessary as an alternative or sequential treatment.

Further Reading

BREAY, H. T. (1982), 'New post-emergence herbicides for grass weed control in sugar beet. . .', *Proc. 1982 Br. Crop Prot. Conf. – Weeds*, pp. 843–8.
 Up-to-date treatment of the potential role of the graminicides in sugar beet.

CROMACK, H. T. H. and DAVIES, W. I. C. (1982), 'Developments in weed control in swedes', *Proc. 1982 Br. Crop Prot. Conf. – Weeds*, pp. 931–8.
 Useful review of recent trends in chemical weed control in this somewhat neglected crop.

JONES, A. G. and ORSON, J. H. (1982), 'The control of grass weeds in annual and perennial crops in the United Kingdom', *Proc. 1982 Br. Crop Prot. Conf. – Weeds*, pp. 793–802.
'State of the art' review of the applicability of the graminicides in a wide range of agric. and hortic. crops.

VARIOUS AUTHORS (1982 (a)), 'Weed control and weed control strategy in arable crops', *Proc. 1982 Br. Crop Prot. Conf. – Weeds*, Session 3A, pp. 55–96.
 A series of papers dealing with two main current issues in sugar beet – the biology and control of weed beet and the 'little and often' approach to weed control in sugar beet.

VARIOUS AUTHORS (1982 (b)), 'Herbicides: the repeat low-dose method', *Br. Sugar Beet Review, 50*, Spring 1982, pp. 7–28.

Very useful series of short articles dealing with various aspects of the low dose/low volume approach to sugar beet spraying.

9 Weed control in grassland

9.1 Introduction

This is a complex area, comprising three main aspects. These are:
(i) the control of weed seedlings in newly-sown grass or grass-legume mixtures;
(ii) weeds in established leys and permanent pastures, including species which have a demonstrable effect on herbage or animal production and also indigenous grasses which may pose a weed problem;
(iii) a relatively small number of problem species, generally either toxic and/or posing special difficulties of control (e.g. ragwort, rushes, gorse and broom); bracken especially could be said to fulfil both sets of criteria.

9.2 Newly-sown grassland

9.2.1 WEED PROBLEMS

Where this forms part of an arable rotation the weeds comprise mainly annual dicots., including groundsel, fat hen, redshank, spurrey and especially chickweed; annual meadow-grass is another very frequent component, in autumn-sown leys especially.

The effect of these weeds on herbage production can be severe. For example, control of heavy infestations of chickweed and annual meadow-grass produced yield responses of up to six-fold in the first harvest following herbicide application (Hagger and Kirkham, 1981). The outcome of this competition can be materially affected by the sowing rate and vigour of the sown grasses; Italian ryegrass is especially valuable in this respect, while timothy and meadow fescue are less so.

Grass reseeds established immediately after grass are more likely to have perennials such as buttercups, docks and rushes and also annuals, including charlock, runch and poppies, whose seeds can withstand prolonged burial. In ploughing older grassland, care should be taken to bury the turf as thoroughly as possible, so as to minimise regeneration of perennial broad-leaved weeds and less desirable grasses.

9.2.2 WEED CONTROL

Cultural methods of control are limited to early grazing or mowing of tall-growing weeds, such as fat hen or redshank. Their success depends on the extent to which the sown species are subsequently able to dominate; low-growing weeds such as chickweed, speedwells and knotgrass may actually benefit from mowing. Chemical control is therefore likely to be required in most cases, summer-sown Italian ryegrass perhaps being the exception.

Generally speaking, spraying should be carried out at the earliest opportunity commensurate with the safety of the sown species, although later-developing perennial weeds may require further treatment. Particularly where a legume is present, the safety of all components of the mixture must be taken into account. Clover seedlings are generally susceptible to herbicide damage until they have passed the first true leaf (unifoliate or 'spade' leaf) stage (Fig. 9.1). Thereafter their tolerance increases with age and herbicide dose should be adjusted accordingly.

Herbicides available for general weed control at the pre-emergence stage include the contact materials paraquat and ioxynil/bromoxynil. The latter can be used in both direct-sown and undersown crops, provided in the latter case that the cereal has reached the three-leaf stage and the grasses and legumes have not emerged. The mixture gives good control of chickweed and mayweeds, but not of cleavers or spurrey. Methabenzthiazuron and ethofumesate are available for grass weed control in perennial ryegrass and ryegrass/tall fescue respectively; both also control chickweed while the former gives some control of other dicots., including mayweeds and speedwells.

Post-emergence control is based on MCPA, 2,4-D and other growth-regulators. Mecoprop is particularly active against spurrey and chickweed, dichlorprop against the polygonums, and dicamba and 2,3,6 TBA against larger weeds generally. Application of any of these herbicides or mixtures is likely to cause damage to the grass seedlings until at least the two to three-leaf stage and for maximum safety it is best to wait until tillering has started before spraying. This may occur anything from three to seven weeks after sowing, ryegrass being most rapid and the small-seeded species, such as timothy, slowest.

Where clovers are present, either in direct or in undersown situations, the butyric acid derivatives MCPB and 2,4-DB should be used. These are converted by dicot. weeds, but not by clover, to MCPA and 2,4-D respectively; 2,4-DB is especially active against polygonums. The translocated material benazolin or the contact-acting bentazone and dinoseb-acetate are also available, the latter singly and the others only in mixtures with MCPB or 2,4-DB; a little MCPA or 2,4-D is often included to improve weed control, at minimal crop risk.

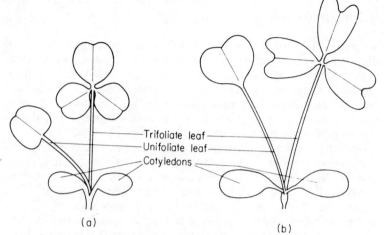

Fig. 9.1: *Legume seedlings at the first trifoliate leaf stage; (a) clover; (b) lucerne. (after Fryer and Makepeace, 1978).*

In grass after grass situations, the main alternative is to kill the weeds in the preceeding sward, and of course direct-drilling of new grass into chemically killed pasture has been a standard technique for many years. The herbicides available include paraquat, dalapon and glyphosate, of which the latter is most expensive, but probably gives the best overall kill. It is necessary to protect the new sward from disease and pest, especially slug damage, and to ensure that it has adequate nitrogen. Any other measures to promote crop vigour, in particular drainage and basic fertility, should be attended to prior to sowing (M.A.F.F. Booklet 2049).

9.3 Established grassland

9.3.1 WEED PROBLEMS

The definition of what constitutes weeds in this situation is less clear. Three main categories can be recognised, however:

(i) *Unproductive species*, which take up space which could be occupied by more productive types. These include creeping or rosette-forming species such as yarrow, dandelion, plantains and docks, and also a range of grasses from couch and annual meadow-grass at one end of the scale to those such as rough-stalked meadow-grass at the other. Rushes and some other monocotyledons could also be included in this category.

(ii) *Noxious species*, including those which are not only unpalatable in themselves, but which restrict grazing in their immediate vicinity and may also reduce the value of conserved fodder, e.g. tufted hair-grass and thistles. Plants which can cause physical injury to livestock or taint animal produce, such as wild barley and wild onion, also fall into this group.

(iii) *Poisonous plants.* This covers a wide range from species whose effects are relatively mild or 'chronic' – i.e. long-lasting but rarely fatal (e.g. buttercup, horsetails) – to lethal types such as yew, water dropwort and cowbane. The two most important species, however, are undoubtedly ragwort and bracken, both of which can show chronic and acute symptoms.

Although all these species are widespread in grassland, there have been few attempts to quantify their occurrence. The main recent source of information is the National Farm Survey (Peel and Hopkins, 1980), covering 500 farms in England and Wales with at least 50% of their area in permanent grass. The most important species, as indicated by the responses of farmers, were thistles (especially creeping thistle) and docks, a long way ahead of nettles, buttercups, weed grasses (including couch, annual meadow-grass and soft brome), and rushes (Table 9.1).

Creeping thistle is especially associated with older swards on freely-drained soils and with low levels of nitrogen and phosphate, i.e. essentially extensively run beef and sheep farms. Docks on the other hand are most abundant on younger swards in mainly dairying areas, with relatively high levels of N and P. 'Significant' infestations of creeping thistle are said to occur on as much as 400,000ha and of docks on 200,000ha; inclusion of low and incipient infestations raises the totals to 1,000,000 and 500,000ha respectively. The situation is broadly similar in Scotland, where creeping thistle is the major dicot. weed in the east of the country, infesting 50% of swards over 10 years old (Swift, 1978), while in the wetter west buttercups and rushes are particularly abundant.

Peel and Hopkins recognise two main sorts of grass weeds: undesirable species including couch, annual meadow-grass, brome grasses, wild barley, creeping soft-grass and tufted hairgrass; and discarded agricultural species such as rough-stalked meadow-grass, red fescue and crested dogstail, which were at one time quite widely included in seed mixtures as 'bottom-grasses'. Together with common and creeping bents, the latter dominate more than half the lowland grassland of England and Wales.

A degree of controversy surrounds these grasses. In the sub-optimal conditions of pH, N, P and drainage which occur in much of the grassland area, their presence evidently does not detract from either pasture or animal production and according to Snaydon (1978) as much as 95% of the variation in these swards is due to environmental and management factors and only 5% to species differences. The opposite point of view, as expressed by O'Keefe (1982), is that the indigenous grasses are intrinsically poorer than perennial ryegrass in yield potential, palatability and digestibility.

Table 9.1: *Relative importance of weeds in grassland, as perceived by* *farmers*

Weed species	Proportion of farmers mentioning it as a problem (%)
Thistles	50
Docks	40
Nettles	17
Buttercups	10
Grass weeds	9
Rushes	6
Chickweed	5
Ragwort	4
Redshank/Fat hen	2
Bracken	1
Others	5

(After Peel and Hopkins, 1980)

All sown swards of course deteriorate with increasing age, at a rate governed both by management-dependent parameters (stocking-rate, grazing regime, fertiliser use) and also by more basic soil and climatic effects. Factors affecting the extent to which weeds invade sown grassland include the following (Roberts, 1982):

  (i) drainage
  (ii) soil pH
  (iii) soil nutrients
  (iv) physical damage
  (v) dung, urine and slurry
  (vi) defoliation
  (vii) winter-kill

The relevance of these is discussed in relation to a number of problem weeds of grassland in 9.4 (below).

9.3.2 WEED CONTROL

The National Farm Survey indicated that only two-thirds of farmers took noticeable action against weeds in established grassland. This mainly involved docks, where spraying was the main tactic, and thistles, where cutting was favoured. Many dicot. weeds are readily controlled by herbicides, but these should not be regarded as a replacement for more fundamental measures to rectify underlying problems of soil fertility, drainage and so on. In purely cost/benefit terms little is to be gained from attempts to control these weeds unless steps are also taken to minimise reinfestation.

Herbicides are, however, the only feasible direct method of control in most cases. Of the alternatives, pulling or digging out weeds is only practicable on the smallest scale, and the former especially is unlikely to lead to lasting improvement. Cutting is limited in its application to species such as creeping thistle, which invest much of their metabolic activity in the production of aerial stems; in ragwort, for example, it may actually be harmful, in inducing the plants to grow on from year to year in a more prostrate, perennial form.

For most broad-leaved weeds the hormone-type weedkillers form the main line of attack, either singly or in mixtures according to the range of weeds involved. MCPA and 2,4-D give adequate control of a number of more or less susceptible species, including thistles, plaintains, daisy and buttercups. For more resistant types such as docks, mecoprop and dicamba mixtures are preferred, although more than one dose may be required. Where it is intended to retain a proportion of clover in the sward it will be necessary to use MCPB or 2,4-DB, alone or in mixtures with benazolin, bentazone or dinoseb. Alternatively asulam may be used against docks, but sown grasses other than perennial ryegrass may be checked or killed; two applications are again usual.

The optimal time of spraying also varies according to species. Generally speaking the weeds must be in active growth and have adequate amounts of leaf if translocation is to be sufficient to kill especially those with protected underground parts. The flower-bud stage has been found to give good results in many cases, while in others spraying regrowth after grazing or mowing may be preferable. Specific restrictions apply to the entry of livestock into the treated area; this is particularly important where poisonous weeds such as ragwort are present, as these may become more palatable to stock after spraying.

A good deal of work has gone into examining the possibility of selectively controlling less desirable grass species in established swards, with rather limited results. The most widely used method involves the use of a low dose of dalapon to suppress creeping bent, rough-stalked meadow-grass and Yorkshire fog in perennial ryegrass or cocksfoot, but not timothy; red fescue and tufted hairgrass are relatively resistant, as is white clover. The sward as a whole is likely to be checked and recovery will depend on the proportion of desirable to less desirable species and on subsequent management. Ethofumesate may be similarly used in all major grasses, including timothy, against meadow-grasses, brome-grasses and wild barley, but clover will be severely checked or killed.

Finally, promising results have recently been obtained from the use of rope-wick and similar appliances: grassland weeds, with their mainly upright flowering stems, are well suited to control by this means. Two passes at right angles may be necessary and provided the equipment is properly used there is no danger of damage to the grass or clover content of the sward. It has been estimated that 44% (600,000 ha) of lowland grassland in England and Wales would be suitable for this technique, having a preponderance of tall-growing weeds, especially creeping thistle, docks and nettles (Oswald, 1982). These are mainly grazing fields, where the height differential would be sufficient to permit the safe use of glyphosate or other chemicals; this was not generally found to be the case in hay or silage.

9.4 Problem weeds in grassland

9.4.1 THISTLES (MAINLY CIRSIUM SPP.)

As indicated, these are the most widespread weeds of lowland grassland, especially on older swards. Spear thistle, marsh thistle, musk thistle and welted thistle are more or less confined to grassland, the last two on calcareous soils, mainly in the south; creeping thistle, easily the most important overall, also occurs on arable land, as do the related sow-thistles.

Creeping thistle (*Cirsium arvense*) is a perennial, sending up leafy and flowering shoots from an extensive, creeping, fleshy root system lying at a depth of between 15″ and 24″ in the soil. The aerial parts die back in autumn and the plant overwinters on the food stored in the roots, which can persist for many years. New shoots develop in the spring, some of which eventually produce flowers in clusters of from two to six small flowerheads. Marsh thistle is similar, but more branching, with spiny stems and an overall purple tinge.

Welted thistle, spear thistle and musk thistle are biennials and spread only by seed. Spear thistle (*Cirsium vulgare*) has a wide distribution and is (like creeping thistle) a scheduled weed under the 1959 Act. The sow-thistles (*Sonchus* spp.) are generally softer plants with more numerous, small flowerheads. There are two annual species, smooth sow-thistle and prickly sow-thistle, and one perennial, field sow-thistle, which has a root system not unlike that of creeping thistle.

Creeping thistle is associated with generally low levels of phosphate and nitrogen but high potassium and is more prevalent on well-drained soils. It appears to be favoured by over-grazing in winter and early spring (which reduces grass competition and creates bare patches suitable for colonization), coupled with under-utilisation later in the year, when the thistles are growing strongly. Improved management and increased fertility are, therefore, an essential background to more specific control measures, which include cutting and spraying.

To be successful cutting must be carried out at least twice during the growing season over a period of years, and spraying will generally be preferred, especially where regeneration occurs after re-seeding. Creeping thistle is susceptible to MCPA and 2,4-D, which will give good control in the year of spraying; follow-up treatment may be necessary in subsequent years if well-established plants are involved. Where clover is present herbicides containing mainly MCPB or 2,4-DB must of course be used.

9.4.2 DOCKS (*RUMEX* SPP.)

Two main species are involved: curled dock (*Rumex crispus*), a biennial or short-lived perennial; and broad-leaved dock (*R. obtusifolius*), a long-lived perennial with a well-developed fleshy tap-root. Fertile hybrids are also common. Regeneration takes place readily from cut plants or from fragments of tap-root resulting from trampling or cultivation, while longer-distance dispersal depends on the abundant seed. This is hooked and is spread by attachment to animals or birds and also by being eaten and passed out in faeces. Manure, hay and straw can also contain viable seed, while water-movement in flooded fields is a further means of localised spread. Seed can germinate soon after being shed, but if buried can survive for many years in the soil. It is important therefore that flowering should be prevented as far as possible.

Docks can grow on a wide range of soil-types, but are clearly associated with intensive situations, often devoted to dairying and involving high levels of nitrogen. The rather open swards produced by rye-grass-only mixtures provide ample opportunity for invasion by docks, and the problem is exacerbated by poaching and uneven or excessive slurry application. In general docks appear to be a severe problem on between 5% and 10% of lowland grassland.

Control by digging or 'spudding-out' is too time-consuming on all but the smallest scale, while cutting or hand-pulling are unlikely to achieve complete kill, especially of established plants. Even cultivations such as discing or rotavation on their own are as likely to spread as to control docks and are probably best used as a prelude to spraying, which is generally more effective on regrowth than on mature plants.

Seedling docks are readily controlled along with other MCPA-susceptible weeds in newly-sown grass leys, as are *small* plants regenerating from root fragments. In all cases, however plants rapidly become more resistant and by the time they are two or three years old repeat spraying with more active materials will usually be necessary for lasting effect. Mixtures of MCPA, mecoprop and dicamba have been successful in these circumstances, although clover is severely checked. Spraying is best carried out when the dock plants have adequate amounts of foliage and are growing (and therefore translocating) rapidly; a good time may be before a hay or silage cut or in the aftermath.

Asulam was introduced primarily for dock control, but though it is successful and clover-safe, timing and weather conditions are more critical than for the alternatives and repeat spraying within prescribed time limits is usually necessary. Timothy, meadow-grasses, bent-grasses and cocksfoot are likely to be checked, the last severely so.

9.4.3 BUTTERCUPS (*RANUNCULUS* SPP.)

The three species of buttercup involved differ both in their biology and in their response to herbicides. Creeping buttercup (*Ranunculus repens*), as the name suggests, spreads by stolons, which root at the nodes and eventually give rise to independent daughter-plants. It is usually associated with impeded drainage and is the most likely to invade short-term grass. It is also the most susceptible species to MCPA or 2,4–D and is best sprayed in spring or early summer, prior to flowering. Both this and the next species have three palmately-lobed 'leaflets', the terminal one being on a short stalk.

Bulbous buttercup (*R. bulbosus*), which can be distinguished by its swollen stem-base and reflexed sepals, is more often associated with old, well-drained pastures and is the species most resistant to herbicides. It flowers early and then dies back for the year, so that spraying is best carried out in the autumn, when the new leaves appear. Meadow buttercup (*R. acris*) has no specialised overwintering structures and differs also in its basic leaf-shape, with three much more deeply dissected 'leaflets' all arising from one point. It is intermediate in its soil preference and in its susceptibility to herbicides, MCPA being more effective than 2,4–D.

9.4.4 RUSHES (*JUNCUS* SPP.)

Common rush (comprising the species-pair *J. effusus* and *J. conglomeratus*) is more or less universal in upland areas and also widespread in the lowlands, wherever field drainage is less than adequate. Another predominantly lowland species is the hard rush (*J. inflexus*); this has caused fatalities in livestock, which may develop a craving for it (Forsythe, 1979). Jointed rush (*J. articulatus*), sharp-flowered rush (*J. acutiflorus*) and heath rush (*J. squarrosus*) are common in the wetter uplands.

Of the above, generally only common rush is sufficiently trouble-some to warrant specific control measures; several of the other species are in any case resistant to herbicides. Spread of the common rush is by the extension of existing tussocks and by seed, which is produced in great quantity and dispersed by a variety of means including water, from which it is protected by an oily outer covering. Spread is encouraged by over-grazing and poaching, both of which expose readily colonised soil. The seeds are particularly long-lived and infestations often arise when old pastures are ploughed.

After ploughing rush-infested land, therefore, steps must be taken to produce a vigorous grass sward, backed up by spraying with suitable herbicides; even then, unless there are radical improvements in drainage, the benefits are likely to be short-lived. Spraying should be carried out with MCPA or 2,4–D in late spring or early summer when the rushes have new growth, but before flowering starts. Cutting the sprayed rushes helps to complete the kill and promotes sward regrowth. In practice, complete control is unlikely in one year and repeat treatments will almost certainly be required.

9.4.5 GORSE AND BROOM (*ULEX* SPP. and *SAROTHAMNUS SCOPARIUS*)

These are short-lived perennial shrubs occurring widely on upland and some lowland (especially coastal) areas, generally on freely-draining, slightly acid soils, and often on quite steep slopes. Dense infestations can occupy useful grazing land to the detriment of grass and stock alike. The seeds are produced in small pods and can survive for long periods in the soil; regeneration can also take place from the stem-bases and rootstocks of old or burned plants. Burning as a control measure is therefore of limited value, unless followed up by grazing or spraying to control regrowth. Alternatively flail-type or chain-saw implements can be used to break up and destroy younger bushes especially, while successful control has been obtained in south-west Scotland with a heavy-duty rotavator.

The recommended chemical control for these and other woody weeds is 2,4,5–T, alone or with 2,4–D; spraying is best carried out in summer and maximum coverage should be ensured. Spray drift must be avoided, especially adjacent to susceptible crops (such as swedes) and forestry plantations. A newer alternative which lacks the over-tones associated with 2,4,5–T and which is reported to be if anything more effective than the former on gorse and broom is triclopyr. Problems can arise because of the steep and broken ground on which gorse and broom often grow, and it may be necessary to use industrial-type knapsack sprayers or hand-lances in order to reach the less accessible bushes.

Application by spinning-disc requires extra care and full protective clothing must be worn. Finally, aerial application is possible in suitable circumstances, although specific clearance must be obtained in each case and particular care must be taken over spray-drift. Aerial application is of course expensive, but may be the only feasible method with large-scale infestations or on more difficult ground. In any case, better results are likely from spraying young regrowth after cutting or burning than from any treatment of mature bushes.

Regeneration is likely, and steps must be taken to encourage a vigorous grass sward by the addition of lime and fertiliser, increased stocking rate and improved grazing management. Where appropriate, cultivation and reseeding may be the best appproach. On the face of it the cost of such operations may seem out of proportion to the value of the land concerned, but where dense infestation occurs they can be seen as a method of acquiring extra grazing at relatively low cost.

9.4.6 BRACKEN (*PTERIDIUM AQUILINUM*)

Bracken, brake or fern is the most pernicious weed of the drier uplands and unenclosed hill land. It now infests some 200,000 ha in Scotland alone and equivalent areas elsewhere in the north of England, the Lake District, Wales and the south west. Much of this spread has taken place in the last hundred years, associated perhaps with a decline in cattle numbers and with general population trends in these areas.

The plant spreads mainly by an aggressive rhizome system, which can extend the periphery of a colony by several feet in a year. Bracken occupies what is often the best soil in the areas where it occurs, generally on moderate, relatively well-drained slopes. It interferes with grass production, makes shepherding more difficult, conceals sickly sheep and is also potentially toxic to all classes of livestock.

Where the terrain allows, ploughing and re-seeding is the preferred method of control, with a pioneer crop of rape, Italian ryegrass or stubble-turnips followed by a suitable cereal or grass mixture. For various reasons, however, this is only feasible on a small proportion of the bracken ground. Where the bracken is sufficiently sparse for the underlying sward to be readily visible, cutting or bruising the fronds twice per year for a minimum of three years will at least reduce its vigour and pave the way for more lasting improvements. If on the other hand the bracken is dominant, chemical control is likely to give the best results.

The standard herbicide for many years has been asulam, applied at about full frond emergence but before senescence has begun, i.e. from early July in the south and mid-July in the north until mid to late August respectively. Spraying can be by ground-based equipment where feasible, but aerial treatment of extensive infestations may be necessary. Hand-held CDA applicators and knapsack sprayers are suitable for small-scale operations. The effect of the chemical is not usually evident until the following season, when little or no regrowth should occur, but can last for up to three or four years before reinfestation begins to take place.

Most grasses and some dicots, are damaged by asulam and care is necessary, especially with aerial spraying, in order to avoid damage to susceptible crops and other plant-life. Monocots. (e.g. sedges and rushes) and woody shrubs such as heather and bilberry are very little affected. Economic returns from controlling bracken will depend largely on the extent to which the land made available is properly utilised; this may involve 'pump-priming' imputs of lime, slag and fertiliser, leading to increased stock numbers (Williams, 1980). Trampling by stock can significantly suppress regrowth, especially if fencing is used to improve grazing control. If it is intended to proceed to full re-seeding, complete kill of existing vegetation can be achieved through the use of glyphosate; particular care is required, however, with this material, which is not cleared for aerial application.

9.4.7 RAGWORT (*SENECIO SPP.*)

A common, colourful plant of older, usually overgrazed pasture, ragwort is poisonous to cattle and horses both as the growing plant and in hay or silage. Sheep, especially adults, show a considerable degree of immunity and can be used, within limits, to clear up existing infestations and to prevent new ones developing. Common ragwort (*Senecio jacobea*) has a wide distribution mainly on light, sandy soils. It is especially prominent in parts of north-east Scotland, but in England and Wales formed a serious problem in only 1% of fields in the National Farm Survey. The marsh ragwort *S. aquaticus* occurs on heavy or poorly-drained soils and is also most abundant in northern areas.

Both plants are basically biennials, overwintering as a rosette of leaves on the surface of the ground and in the second year producing the characteristic tall stem, with its crowded yellow flower-heads. Trampling or cutting can induce the plant to perennate and to produce flowers in successive years, often with increased vigour. Many seeds are produced and are an important means of spread especially into young grass. Short distance spread also occurs through the development of new shoots from the stem bases or from root fragments. The seeds can remain viable in the soil for several years and can lead to rapid colonisation of bare patches in the sward caused by poaching, over-grazing, winter-kill or the effects of rabbits and moles.

Cutting is therefore not recommended as a means of control, except before ploughing. The same is true of pulling, which in any case is only practicable on a very localised scale. Ragwort is readily killed by ploughing and is not a significant problem in arable situations. Where grass follows grass however, re-infestation from seed or vegetative fragments is a strong possibility and must be countered by adequate fertiliser treatments and good grazing management, including the avoidance of undue poaching. Lasting control of marsh ragwort may require soil drainage, while on particularly infertile or shallow soils even good management may need to be supported by chemical treatments on a regular basis for lasting control.

Control of ragwort can be obtained with either MCPA or 2,4-D, the latter being slightly more reliable. Work in north-east Scotland has led to recommendations which in many cases exceed those given by manufacturers but which provide superior control of ragwort, although clover may be seriously checked . Mixtures of these herbicides with e.g. mecoprop and dicamba also give good control but are even more damaging to the clover and are also relatively expensive. Ragwort is resistant to MCPB and 2,4–DB and there are no fully selective materials currently available to control the weed in clover-rich situations.

Time of spraying is largely governed by the intended purpose of the sward. In pasture the most appropriate time is late April to late May, not at the flower-bud stage as previously recommended. In silage or hay fields on the other hand the preferred time is the preceding autumn, which gives the weeds ample time to disintegrate and removes the serious risk attaching to the presence of ragwort plants in hay or silage, possibly the main source of poisoning. Duration of control is best with spring spraying, but the long-term effect will again depend on subsequent management.

9.4.8 SOFT BROME (*BROMUS MOLLIS*)

Soft brome is an annual or biennial grass which for some years has been spreading in hayfields, mainly in Scotland and the north of England. Its life-cycle is perhaps not fully understood but it appears to germinate either in late summer, overwintering as a seedling or, in Scotland especially, in spring. It is extremely competitive and yield increases of up to 100% have been claimed following its control by herbicide (Harkness and Frame, 1981). Soft brome is favoured, like many other grassland weeds, by winter-grazing, poaching and low fertility.

The main control objective is the prevention of seeding, as the plant seeds profusely and the seeds can remain dormant in the soil for many years. This can be achieved by a switch to silage or at least an altered grazing regime, but neither of these options is feasible in many situations where soft brome occurs and chemical control is therefore often required. Ethofumesate has given very good results, up to complete kill, but is destructive of clover. It is probably best used in late winter, when it controls seedlings by foliar action, while its residual effect also takes out the spring-germinating seeds (Goldsworthy and Drummond, 1981). An alternative is a mixture of TCA and dalapon, which has given almost as good control and does not harm clover, although there is some check to the sward as a whole in the year of application (Cooper, 1982).

9.5 Economics of weed control in grassland

Discussion of this topic is hindered by the diffuse nature of the problem and of its effects on herbage and ultimately livestock production. As indicated earlier, the weeds themselves often have some feeding value and in any overall assessment this has to be balanced against reduced productivity. In grassland, too, the short-term effects of control may be less obvious and the long-term benefits correspondingly more important in assessing the final result, which is also considerably affected by the way in which any increased production is utilised (e.g. to increase stocking rate, or to replace bought-in hay or concentrate).

Evidence as to the actual returns from controlling weeds in grassland, especially newly-sown grass, is extremely limited, but where weeds such as soft-brome, rushes or ragwort are present the use of specific control measures is likely to be fully justified. Doyle (1982), reviewing this situation, notes that to a certain extent the value of increased production depends on when it becomes available in relation to the preferred farming system and points out that, in general, chemical weed control in grassland is likely to be two to three times as expensive as in cereals, relative to the value of the crop. In this situation it only remains to emphasise the absolute necessity of adopting management practices such as will generate maximum results from the sown sward and from any control measures undertaken.

Further Reading

DOYLE, C. J. (1982), 'Economic evaluation of weed control in grassland', *Proc. 1982 Br. Crop Prot. Conf. – Weeds*, pp. 419–27.

Useful coverage of a little examined subject; includes discussion of the impact of weeds on livestock production and of the economic value and cost-effectiveness of control.

M.A.F.F. (1982), 'Ragwort', M.A.F.F. Leaflet No. 280.

Full coverage of biology and new thinking on control of this major poisonous weed.

PEEL, S. and HOPKINS, A. (1980), 'The incidence of weeds in grassland', *Proc. 1980 Br. Crop Prot. Conf. – Weeds*, pp. 877–90.

Definitive contribution on the occurrence and distribution of both grass and dicot. weeds in new and established grassland in England and Wales.

O'KEEFE, M. C. (1982), 'The role of herbicides in increasing energy yields from U.K. grassland', *Proc. 1982 Br. Crop Prot. Conf. – Weeds*, pp. 357–62.

Concise but telling presentation of the case for grassland improvement via improved management, sown species and the application of nitrogen and herbicides.

ROBERTS, H. A. (ed.) (1982), 'Weed Control in Grassland', *Weed Control Handbook*, 7th edn., Blackwell Scientific Publications, Oxford. pp. 351–68.

Useful summary of basic distribution and principles of control, but mainly valuable for treatment of factors affecting weed content of grasslands.

SNAYDON, R. W. (1978), 'Indigenous species in perspective', *Proc. 1978 Br. Crop Prot. Conf. – Weeds*, pp. 905–13.

Presents a well-documented case for the value and potential of existing as against sown grasses – see O'Keefe (1982) for the opposite viewpoint.

10 Fruit and vegetable crops

10.1 Introduction

It is beyond the scope of this book to provide detailed information on every herbicide, herbicide mixture and herbicide programme available to the horticultural industry. Instead, discussion will concentrate on the types of chemicals which may be used in a number of horticultural situations, and on certain principles of weed control, including the use of herbicides, which help to minimise the growth of weeds, and to optimise their control.

In relation to each situation considered, a number of the more common chemicals will be mentioned. This will be by no means an exhaustive list of the herbicides which are available for these situations.

A booklet containing the full range of approved pesticide chemicals, including herbicides, available in the U.K. is published annually by MAFF under the banner of the Agricultural Chemicals Approval Scheme (ACAS). [See Appendix 2] This booklet, entitled *Approved Products for Farmers and Growers* is available from Her Majesty's Stationery Office, and is strongly recommended. Furthermore, MAFF publish a number of free booklets on various aspects of horticultural weed control and these provide an invaluable addition to the available weed control literature.

In addition, herbicide application rates are not generally included as these tend not only to vary from one area to another, but may, indeed, require to be altered to suit soil conditions in different fields on the same holding. Recommended application rates are readily obtained either from the manufacturers' literature or from the MAFF publications mentioned earlier. The ACAS booklet does not, however, provide application rates for chemicals.

No matter what the situation, whether it is a site for a new lawn or a field for raspberries, the motto which should always be adhered to is START CLEAN AND STAY CLEAN. A great deal of time, effort and money can be saved, in the long term, if adequate care is given initially to cleaning and preparing land before planting.

This is especially true in relation to perennial weeds. In the past when cultivation was much more the 'order of the day' in horticulture, perennial weeds presented less of a problem since, with very few exceptions, these weeds are unable to withstand repeated cultivations. On the other hand very few, if any, perennial weeds are satifactorily controlled by soil-applied herbicides at the rates at which they are used selectively in horticultural crops and amenity plantings.

Annual weeds do not present the same problem since virtually all species are controlled by the judicious use of soil-applied chemicals.

It is important, therefore, that every effort is made to get rid of perennial weeds especially on land to be cropped, which means making maximum use of foliage-applied, translocated herbicides. Chemicals such as glyphosate, aminotriazole and dalapon are particularly useful for this purpose, although dalapon is used primarily for the control of grass weeds. Paraquat can also be used but this chemical is generally only effective against annual weeds.

Since paraquat and glyphosate have virtually no persistence in the soil, crops can be planted almost immediately after spraying. However, these two herbicides act in essentially different ways. Paraquat, a contact herbicide, exerts its effect very quickly and the weeds which it controls die in three or four days during the summer months. Weeds die more slowly when sprayed during the winter months, but control is usually no less effective. Glyphosate, on the other hand, being a translocated herbicide, acts much more slowly, and it may take three to four weeks before the plants are completely destroyed.

The different modes of action of paraquat and glyphosate, and the different time intervals required for these two chemicals to achieve plant kill, mean that a slightly different approach must be adopted when it comes to cultivation.

Under normal circumstances, with paraquat the area can be cultivated about five to seven days after spraying. By contrast, glyphosate works best when the root and rhizome system of the plant is intact, as this allows translocation of material throughout the plant. Because of this, cultivation should not take place until leaf symptoms (reddening, then yellowing) appear, which is usually 10–14 days after an autumn spraying. Complete desiccation normally occurs within 30 days, but during cold weather it may be 30–50 days before leaf symptoms are noticeable.

No matter how much care is taken to eradicate perennial weeds before planting there will inevitably be some reappearance of these plants in the crop or shrub bed. In the first few years after the start of the herbicide programme these plants may arise from seeds or from root fragments or rhizomes that have survived the previous treatment but, generally, they can be controlled by spot treatment. This will be discussed in detail in relation to individual situations.

Once a crop has been sown or planted, or an area has been landscaped, the most important weapons in the horticulturist's armoury are soil-applied herbicides. By careful choice and correct use of these chemicals, whether alone, in mixtures, or in herbicide programmes, the horticulturist can be virtually assured of efficient control of annual weeds in almost every situation.

There are, however, odd occasions when these chemicals either do not provide satisfactory weed control or may cause injury to non-target plants.

Generally, unsatisfactory weed control results either because the herbicide was applied at the wrong time or because a period of dry weather followed application. To be effective, soil-applied herbicides have to be taken up by weed roots, but if annual weeds have become established their roots are usually located at a greater depth in the soil than the zone where most soil-applied herbicides would be found (i.e. the top 5cm). This is why soil-applied herbicides are ineffective if they are applied too late.

During a period of prolonged dry weather soil-applied herbicides may be ineffectve because rain is usually required to wash these chemicals from the soil surface down to the region where the weed seeds are germinating. Under dry conditions the weeds grow un-harmed through the herbicide layer, which remains on the soil surface.

The latter problem may, to some extent, be overcome either by spraying, if appropriate, in the late winter or early spring before the soil dries out, or by irrigating before spraying to guarantee that the soil will be moist. In certain circumstances irrigation also serves to avoid herbicide injury to newly-planted crops as it helps to consolidate soil around the roots.

Injury to crops or ornamental plants may be caused in a number of ways. If a period of very heavy rain follows treatment, crop damage may result from herbicides being washed deeper into the soil than normally would be the case. This is especially true on light sandy soils.

In addition, injury may be caused by over-dosing as a result of errors in calculation of dose rate, inaccurate calibration of the sprayer, faulty nozzles, overlapping swaths, and spraying a strip twice because of not marking the finishing point before leaving the field to refill the sprayer.

Whereas under-dosing generally only leads to unsatisfactory weed control and may be remedied by re-spraying, over-dosing can have quite disastrous effects on the crop.

For soil-applied herbicides to be effective without causing damage, the following points should be kept in mind:

a) read the label – all herbicides must be applied at the recom-mended rate and at the recommended time.
b) the weeds must be properly identified.
c) herbicides should be applied to moist, weed-free soil, preferably when rain is forecast.
d) soil-applied herbicides only work properly when rain-water carries them down into the soil to the region of germinating weed seeds (i.e. the top 5cm).

e) soil-acting herbicides should be applied to soil of a fine, firm tilth, since they will not penetrate below large clumps of soil where weed seeds can germinate.
f) always check calculation of dose rate and calibration of sprayer, and make sure the nozzles are working properly.
g) make sure the spray operator is properly trained in the required procedures.

Having dealt with a few general, but important, principles let us now consider the control of weeds in some horticultural situations.

10.2 Top fruit

The majority of top fruit orchards still utilise the system whereby the trees are grown in strips treated with herbicides, with grass between the rows of trees. This practice started just after the First World War and resulted, in the main, from the development of gang-mowers which were used on the airfields during the war. This system of grass alleys provided farmers with an alternative to cultivation and its associated problems concerning the movement of vehicles under adverse weather conditions.

Present evidence, however, suggests not only that production of fruit of better size and quality results from the elimination of weed growth, but that the removal of grass alleys is also important in this respect. In addition to competing with the fruit trees for moisture and nutrients, grass alleys also make weed control more diffcult, since they provide a source of stoloniferous and rhizomatous weeds, and can contribute large numbers of weed seeds to the herbicide-treated strips.

At present, the indications are that the removal of all vegetation from orchards (weeds and grass) leads to increased growth and yield of fruit. This fact was not appreciated during the early days of the grass alleys since the main type of pruning used at that time was spur pruning. This involved the removal of most of the new growth, so that any reduction in growth caused by the presence of the grass was not detected.

Despite the weight of evidence in favour of total herbicide managment, growers are reluctant to move over completely to this system. This may stem from anxiety regarding possible problems of rutting and lack of traction in the absence of grass. In addition, conventional spray booms and nozzles are generally unsuitable for orchard spraying. However, the development of new spraying systems and nozzles which do not require a boom have, to an extent, overcome this problem.

There are undoubted advantages in adopting overall herbicide application since there is evidence that a better economic return can be expected from this system. Moreover, complete removal of vegetation may in certain instances give a reduction in tractor wear and fuel consumption, and release labour for other important work. Furthermore, the advent of new spraying systems and nozzles which reduce the risk of drift means that weeds can be sprayed fairly quickly under most conditions, and at the seedling stage when they are most susceptible. Overall herbicide systems are unsuitable, however, on steep slopes where soil erosion is likely to be severe.

10.2.1 APPLES

Although soil-applied herbicides form the basis of weed control programmes in orchards, the removal of perennial weeds before planting is also important and this normally involves using herbicides such as glyphosate, 2,4–D and aminotriazole.

At the dose at which herbicides are used in crops, and with the vast array of weeds which are present in orchards, no one soil-applied herbicide is capable of controlling all the weeds. Nowadays, therefore, increasing use is being made of herbicide 'cocktails' which normally consist of two or, occasionally, more herbicides. These cocktails usually contain combinations of aminotriazole, 2,4–D, mecoprop, MCPA, plus soil-applied materials such as simazine and diuron. The use of herbicide mixtures provides a much more efficient method of weed control as the range of weeds on which they exert an effect is greatly increased.

It is important to appreciate that not all chemicals can be combined in cocktails as they may not be compatible either biologically or chemically. If they are not compatible then they may become ineffective, resulting in unsatisfactory weed control. Also, incompatibility may lead to loss of the selectivity which each chemical separately possesses, and this can have disastrous results on the crop.

It is essential, therefore, that growers do not formulate their own cocktails. Instead they should use proprietary mixtures or, alternatively, seek professional advice.

For problem weeds such as perennial ryegrass (*Lolium perenne*) and annual meadow-grass (*Poa annua*), which tend to be resistant to many of the commonly-used herbicides, many growers are turning to alternative materials such as glyphosate. This chemical is currently recommended as a directed spray in established apples and pears between leaf fall and green cluster, but growers are tending to make much wider use of this material. For instance, some are treating problem perennial weeds with glyphosate by means of the Croptex Herbicide Glove. Others are using the chemical for 'spot treatment', while some are applying it as a directed spray in orchards even when the trees are in leaf.

Although paraquat is still commonly employed in orchards to control overwintered weeds, there is a growing preference for low doses of aminotriazole which will check many perennial weeds as well as control annuals.

If it were not for the fact that many growers consider chlorthiamid and dichlobenil to be too costly, these herbicides would provide good control of emerged annuals, as well as certain perennials that are difficult to control such as horsetail (*Equisetum* spp.) and coltsfoot (*Tussilago farfara*).

Growth-regulator herbicides like 2,4–D and MCPA give good control of perennials such as stinging nettle (*Urtica dioica*), field bindweed (*Convolvulus arvensis*) and dock (*Rumex* spp.). These herbicides may be applied as directed sprays around the base of fruit trees (not during the blossom period), or as spot applications elsewhere in the plantation. However, all fruit trees are sensitive to these chemicals, and it is essential to ensure that there is no spray drift. Moreover, since ester formulations of these chemicals are extremely volatile and the vapour can damage the trees, they should not be used in orchards.

Although there are a number of herbicides available for use in established apples and pears, the situation in newly-planted crops poses a greater problem. The fact that the market is fairly small, and the risk of crop damage greater than with well-established trees, has meant that herbicide manufacturers have not been forthcoming with recommendations for newly-planted crops. Where soil-applied herbicides have been tried, growers' fears about crop damage have occasionally been substantiated by the appearance of yellowing on leaf margins.

An alternative worth considering in a newly-planted crop is the use of a black polythene mulch. Crops in which these mulches have been used show increased growth compared with those in which herbicides were used, or which were hand-weeded. This increased growth is attributed principally to an increased availability of soil moisture, although an increased nutrient supply may also play a part. This method also produces increased growth in newly-planted trees when compared with straw mulching.

While there may be problems involved in laying the mulch over a large area, consideration should be given to trying this method at least on a small trial plot in the orchard. Initially there need be no great expense involved since, in the past, old fertiliser bags and other such materials have been used for this purpose.

10.2.2 PEARS

The reduction in the number of pear orchards with grass alleyways seems to reflect the view of most growers that this system produces a lower yield and smaller fruit than overall herbicide treatment or cultivation, because of competition from the grass, especially for moisture.

With the exception of terbacil and dicamba mixtures, the herbicides available for use in pear orchards are the same as those for apples. However, pears appear to be less tolerant of the chemicals and great care is required in their use.

10.2.3 STONE FRUITS

The use of herbicides in stone fruits is restricted since, generally, these crops are much less tolerant of chemicals, especially soil-applied materials. However, the indications are that where soil-applied chemicals are used, there is an overall improvement in crop performance. Application of low doses of simazine and propyzamide two to three times a year gives good control of germinating weeds, while established seedlings can be dealt with by paraquat or aminotriazole.

10.3 Strawberries

It is no exaggeration to say that the yield and profitability of modern strawberry production is closely related to the efficiency of the weed-control measures employed.

In comparison with other fruit crops, weed control in strawberries is much more difficult to achieve, mainly because of the crop's characteristics. For instance the crop plants are similar in size and form to many weeds, and the crop may be planted over several months, which demands a great deal of flexibility on the part of the herbicides used. In addition, there may be a wide variation in the age and number of crop plants present. Combine these features with the usual problems associated with the use of soil-applied herbicides, such as variable soil and climatic conditions, and it becomes much more difficult to achieve safe, efficient weed control.

In the past, weed control in strawberries relied heavily on cultivation, which undoubtedly had an adverse effect on the crop resulting from disturbance of the soil and associated damage to the crop roots. Moreover, present evidence suggests that, in the absence of weeds, the growth and yield of strawberries is as good, if not better, under zero cultivation, as in management systems utilising cultivations.

That being said, however, if after the crop is planted large clods of soil are left, then a light cultivation is advisable in order to produce conditions suitable for strawing down and picking, and to eliminate large lumps of earth which encourage weed growth.

The most efficient method of controlling weeds in strawberries nowadays is to make full use of available herbicides before planting, at the time of planting, and in the established crop.

It is vitally important to make an attempt to get rid of perennial weeds prior to planting since their removal at a later date is difficult to achieve without harming the crop plants.

In a high-value crop like strawberries, some growers use the soil sterilant dazomet before planting. Although this is expensive initially, growers who adopt this system are convinced that it more than pays for itself in terms of increased growth and yield, and by a reduction in time spent on controlling weeds at a later stage.

By using foliar-applied translocated herbicides in conjunction with cultivation, good control of perennial weeds can be achieved. For translocated herbicides to achieve maximum effect, weeds should have a reasonable leaf area. It is not always possible to achieve this situation in one year by means of cultivation and, if at all possible, a two-year fallow should be considered.

Glyphosate provides good control of annual weeds and the majority of perennial weeds, while dalapon and aminotriazole are good alternatives where couchgrass (*Agropyron repens*) is the main problem. Field bindweed (*Convolvulus arvensis*), thistle (*Cirsium* sp.) and docks (*Rumex* sp.) are all susceptible to treatment with 2,4–D, but better control of docks may be achieved using asulam.

While there is no necessity to cultivate after glyphosate has been applied, a deep cultivation is required three to four weeks after 2,4–D, aminotriazole and dalapon have been sprayed. Moreover, strawberries should not be planted for two months following aminotriazole application, and this increases to four months in the case of dalapon.

It is inevitable that in spite of every effort to eradicate perennial weeds some will appear, either from seed or from perennating organs, in a newly planted crop. However, seedlings of many of these weeds can be controlled with herbicides such as propachlor, trifluralin, lenacil and chlorthal-dimethyl. If seedlings of perennial weeds are not controlled at this stage, or if perennial weeds are becoming re-established because of the failure of pre-planting control measures, the only follow-up treatments available then are spot-treatment, hoeing or hand-weeding.

In addition, paraquat can be used immediately before planting, so that the ground is free of weeds when soil-acting herbicides are applied.

Although a number of soil-applied herbicides are now available for strawberries, simazine, because it is efficient and relatively inexpensive, is still one of the most commonly used. One problem associated with the use of simazine in newly-planted strawberries is that it may in certain instances cause crop damage. Cases of simazine damage are usually associated with heavy rainfall, and tend to occur soon after planting. Damage in England and Ireland is also often associated with treatment between January and July. This problem in newly-planted strawberries can, however, be overcome to a very great extent by dipping the roots of the strawberry plants in activated charcoal. This can be a fairly unpleasant process, but by mixing the charcoal with water to form a slurry, a certain amount of the unpleasantness can be removed from the task. Provided this process is carried out properly, it should afford a fair degree of protection from simazine and other herbicides, which might be leached more deeply than usual into the soil following heavy rain. No single herbicide will control every weed and, consequently, herbicide mixtures or programmes are required in order to obtain effective control of the wide range of weeds which may be present in the crop. It is important, therefore, to be aware and keep a record of the weeds which are present in any particular situation, so that the correct choice of herbicide mixtures or programmes may be made. The combination of herbicides within a mixture or programme is based on the fact that they each control certain weeds which the others do not and, because of this, a much better degree of weed control within the crop is achieved than would be the case if each herbicide were used on its own.

Each herbicide has its benefits and limitations, and provided the grower is aware of these, then, in general, these chemicals provide effective weed control with little risk of damage to the crop.

Lenacil can be used at any time as long as no other soil-applied herbicide is used for three months afterwards. This chemical is much more effective when applied early on in the season when there is plenty of soil moisture available, usually early February. Later in the season when the soil tends to be drier, it is virtually inactive. After February spraying, it may be advisable to apply a 'topping-up' treatment of three-quarters of the normal dose just before the crop comes into flower, in order to maintain satisfactory weed control during cropping. This is usually quite safe even though it does not follow the three-month rule mentioned previously.

Trifluralin is another commonly-used herbicide for newly-planted strawberries, and is normally applied during the two weeks before planting. It is more effective in dry conditions than lenacil, and it controls speedwell (*Veronica* sp.) and field pansy (*Viola arvensis*), which lenacil does not do, but it must be incorporated within two hours of spraying, and an additional herbicide is required after planting to deal with weeds resistant to the trifluralin.

Other soil-applied herbicides of value in strawberries include ethofumesate, propachlor and pendimethalin, all of which give good weed control and are relatively safe to the crop. Ethofumesate which, at present, is recommended only for use on Cambridge Favourite, gives particularly good control of chickweed (*Stellaria media*). It also inhibits clover (*Trifolium* sp.), cleavers (*Galium aparine*) and annual grasses. Propachlor, widely used in vegetables following pre-planting application of trifluralin, also seems to be safe in strawberries, giving good control of many annual weeds, especially cleavers, although it does not control knotgrass (*Polygonum aviculare*) or redshank (*Polygonum persicaria*).

Pendimethalin, a soil-applied residual herbicide for annual weed control in winter barley, provides the strawberry grower with what seems to be a very useful material. It is more successful in controlling weeds in dry weather than other soil-applied chemicals, and is effective against the majority of annual weeds including cleavers, field pansy and speedwells, which are not controlled by some of the standard strawberry herbicides. Groundsel, however, is resistant.

Chlorthal-dimethyl, either alone or in combination with propachlor, gives good control of a wide range of annual weeds, especially speedwells.

In practice, although not recommended, a mixture of low doses of simazine and propyzamide has been found to provide safe and useful short-term control of annual weeds.

No matter what herbicides are used, some weeds will not be controlled. It is important to eradicate these plants before they produce seeds, from which a resistant population of weeds could arise. This can be achieved, to some extent, by using phenmedipham, a foliage-applied contact herbicide, which controls speedwells, chickweed and groundsel as seedlings. Grasses are not controlled however. It is worthwhile considering a mixture of phenmedipham and barban as this gives better control of the seedling stage of black bindweed (*Polygonum convolvulus*), knotgrass and redshank than phenmedipham alone.

Normally there is less risk of crop damage from soil-applied herbicides where the soil contains a reasonable amount of organic matter, and on heavy clay soils. Restrictions on the use of chemicals with regard to soil type and organic matter content are generally referred to on the container, and should be adhered to closely.

Without doubt the most troublesome weeds encountered in strawberries are perennials, and among the worst offenders are couchgrass, field bindweed, creeping thistle and field horsetail. The appearance on the market of 3,6-dichloropicolinic acid, a translocated herbicide, has been a great boost in the control of creeping thistle. Although there have been some cases of minor crop damage with this herbicide, these are insignificant when compared with the effects of the uncontrolled growth of creeping thistle.

An alternative used by some growers to control creeping thistle, is to mix glyphosate with wallpaper paste in a 1:4 ratio and to 'spot-treat' this weed using an oil can.

For the control of field bindweed, consideration should be given to the use of 2,4–D amine. This herbicide appears to be reasonably safe on strawberries, especially if applied early in the season. Because under certain circumstances (for instance if it is applied at the wrong time or at the wrong dose) the use of 2,4–D amine can result in the death of strawberry plants, it is probably safer to restrict the application of this herbicide to areas of the crop which are especially weedy.

Most of the time soil-applied herbicides work well and do not cause crop damage, and there are certain steps which can be taken to try to ensure that this is the case. Sprayers should be checked regularly for faulty nozzles and to confirm that they are calibrated correctly. Herbicides should be carefully chosen and applied strictly according to manufacturers' recommendations. Moreover, there is generally less risk of crop damage if good-sized, healthy plants only are purchased, and care is taken to plant them properly.

An additional aid to weed control, which might be given some thought, is to mulch with black polythene. Not only is this an effective method of controlling annual weeds – it can produce a higher yield and better-shaped fruit. For this system to be successful, it is essential that the ground is prepared properly, cleared of weeds, and that the soil is reasonably moist when the polythene is laid. It is virtually impossible to do anything about this once the polythene is down. Also, the use of a polythene mulch is most effective when coupled with trickle irrigation. The areas between the strips of mulch can be kept free of weeds by herbicides like paraquat and simazine.

Although the reasonably moist soil conditions in the UK mean that there is not a great need to use a polythene mulch to conserve soil moisture, and the initial cost is quite expensive at (currently) about £800-£1200 per hectare, these disadvantages may well be offset by earlier cropping, better yield and quality of fruit, and a saving on the cost of herbicides and the purchase of straw.

The control of unwanted runners usually involves spraying these with paraquat but, occasionally, this causes crop damage. This may happen because the spray comes into contact with the parent plant after careless spraying. Alternatively, the paraquat may be translocated from the runner into the parent plant and this, when it occurs, usually follows autumn spraying under very dry conditions when the parent plants are under drought stress. This damage can be avoided by separating the runners from the parent plants with a slicing disc before spraying, but this treatment may also damage the surface roots of the crop.

An alternative control for runners is dinoseb-in-oil, which acts mainly as a foliage-applied contact herbicide, and there is virtually no translocation out of the runners. Because of this, dinoseb-in-oil is slightly less efficient than paraquat in killing the runners and a number of applications may be required. An advantage of dinoseb, however, is that because there is no translocation it is not necessary to separate the runners from the parent plants.

Both of these chemicals are extremely toxic to man and are governed by the Poisons Rules. In addition, dinoseb comes under the

Health and Safety (Agriculture) (Poisonous Substances) Regulations, which means that full protective clothing must be worn by the operator when applying this chemical.

10.4 Cane Fruit

It is almost impossible to achieve complete eradication of perennial weeds prior to planting cane fruit. However, if maximum use is made of translocated herbicides, in combination with cultivations in the year before the crop is due to be planted, the land can be cleaned up fairly well.

Perennial weeds are not killed either by the presence of a grass ley or by the application of growth-regulator herbicides to cereals, their growth is only suppressed. This means that if cane fruit is to follow cereals or a grass ley, certain perennial weeds like creeping thistle and couchgrass, although they may not be obvious, are likely to be present. Therefore, unless an attempt is made to control these and other perennials in the year prior to planting, they will cause problems later on.

For cane fruit, herbicides like glyphosate, aminotriazole and dalapon give good control of perennial weeds pre-planting. However, it is essential that recommendations concerning ploughing, and the minimum time-interval required between applying the herbicide and planting the crop, are strictly observed.

In raspberries, as an additional measure to control annual weeds, trifluralin can be used any time in the two weeks immediately prior to planting, but one full day, at least, should be allowed between incorporation of this herbicide and planting the canes. This treatment is particularly useful on light, sandy soils where lenacil could cause crop damage.

After the planting of any bush or cane fruit, a residual herbicide should be applied as an overall spray for the control of annual weeds, so that the crop gets a good start. In most cases this means using either lenacil or simazine. In raspberries, however, a number of alternatives are available such as bromacil, chlorthiamid and atrazine + cyanazine. Bromacil also gives good control of couchgrass and stinging nettle (*Urtica dioica*) while, depending on the dose, chlorthiamid is useful against coltsfoot (*Tussilago farfara*), docks, thistles and coughgrass.

For annual weed control in the established crop, a wide variety of soil-applied herbicides is available. These include atrazine alone, or in combination with cyanazine (raspberries only); bromacil; chlorpropham + fenuron (crops established for three years); chlorthiamid (raspberries only); dichlobenil; lenacil; propyzamide and simazine. It is important to remember that every soil-applied herbicide cannot be used in every situation, and any choice of herbicide programme should take into consideration the crop to be treated, cost involved, weeds to be controlled, soil type and herbicides previously used. While dichlobenil and chlorthiamid are very expensive in comparison with herbicides such as simazine, they have the additional advantage of controlling many perennial weeds selectively in cane fruit. One way of reducing the cost of these chemicals, while maintaining the benefits, is to use them by means of band application, whereby the granules are applied only to the crop row.

Where perennial weeds appear in the established crop these may be treated with glyphosate although, at present, the only recommended method of application in fruit crops is by means of the Croptex Herbicide glove. Because glyphosate can translocate over long distances within a plant, there is a danger that spray applied to the inter-row area might be picked up by suckers and moved into the parent plant.

The use of paraquat as a guarded spray during the growing season in raspberries gives good control of most annual weeds, and also checks some perennials especially creeping buttercup (*Ranunculus repens*), which can be eradicated by winter treatment. Paraquat is most effective in conditions where low temperature and light intensity prevail. When mixed with simazine, this combination not only provides successful control of established annual weeds, but also destroys them as they germinate.

Provided the correct choice of herbicides is made (if in doubt, seek advice) and they are applied in the proper manner, there is no reason why fruit crops cannot safely be kept weed-free.

An additional problem in raspberries is the necessity of controlling suckers between the stools, and along the length of the crop row. Crop roots also grow into the alleys between the rows, where they produce suckers which also require to be controlled. Suckers tend to be produced in late March or early April, and if they are not removed they can cause serious reductions in fruit yield.

Hand-hoeing around stools, and rotary cultivation between the crop rows, provide efficient control of suckers, and these do not re-grow as quickly as when contact herbicides are used for their removal. Moreover, in raspberries, any damage to surface roots caused by cultivation does not, apparently, have any significant effect on crop growth or yield. This method is, however, labour intensive, and it may reduce the effectiveness of soil-applied herbicides both by breaking the herbicide 'seal' and by stimulating annual weed seeds to germinate.

The mower and the chain-flail are useful for keeping suckers down between the crop rows, but both have disadvantages. Unless the blades of the mower are set very low, this technique leaves a stubble which makes an uncomfortable environment for pickers. Furthermore, use of the chain-flail can be harmful to young stools and low-growing crop laterals by throwing up stones. By the use of a tractor-mounted rotary pruner set in one position, suckers along the crop row can be controlled. Alternatively, this machine can be adapted to remove suckers growing between the stools by means of a hydraulic arm which positions the pruner between the stools while skirting round the stools themselves.

Nowadays, chemical control of suckers is widespread. Initially paraquat was used for this purpose but it was not altogether successful. On the other hand, dinoseb-in-oil, as a guarded spray, provides a very useful, quick-acting alternative for sucker control, in addition to giving good control of annual weeds.

Additional benefits of the chemical control of suckers include avoiding damage to crops and machinery caused by mechanical control methods on stony land. Moreover, the soil remains undisturbed and, in turn, the herbicide 'seal' remains unbroken. This method also removes the need for 'ridging-up', a practice which can inhibit stool-cane production.

It should be emphasised that dinoseb-in-oil may only be used on raspberries, and it is an extremely toxic chemical for which full protective clothing is required.

If only one soil-applied herbicide is being used certain weeds will remain unaffected, and could soon get out of hand. It is important, therefore, to 'ring the changes' as far as herbicides are concerned, in order that no one species becomes dominant and, consequently, very difficult to control.

10.5 Vegetable crops

Because of the variety of vegetable crops grown, and since there are a large number of herbicides available for use in these crops, in this section only the main principles governing the use of herbicides in vegetable crops will be described. In addition, a few suggestions will be given as to the ways in which choice of herbicide should be made and how the best results may be obtained.

Detailed information, including cost, application rates, herbicide mixtures, and any limitations on the use of individual herbicides (e.g. soil type) may be obtained from manufacturers' literature and from a number of MAFF publications including:

B2258 – Weed control in vegetables
B2251 – Weed control in horticultural brassicas
B2069 – Leeks
B2204 – Carrots
B2206 – Lettuce

Vegetable crops present a number of special problems with regard to weed control. For instance, growing crops at close spacing, either to make full use of available space, or in order to produce crops like carrots and cauliflowers of a restricted, marketable size, means that there is little chance of cultivating to control weeds once the crop is established. Also, the mechanical harvesting of many vegetable crops necessitates the complete eradication of weeds to prevent contamination of the crop, and to avoid obstruction of the machinery. Failure to practise a system of crop rotation with vegetables can lead to severe weed problems, since the same herbicides are being used time after time, and weeds resistant to these chemicals will soon prevail. On the other hand, where a succession of different vegetables are grown there can also be a problem, since soil conditions, the weather and the weed flora will change over the season, and this means that a number of herbicides or herbicide programmes will be required in order to provide satisfactory weed control.

While it may not be possible to overcome entirely the difficulties involved in weed control in vegetables, if the grower adopts the correct strategy they can be greatly reduced.

For instance, since weed competition is most serious in the early stages of crop growth, it may be advisable to employ the 'stale seedbed' technique. This involves cultivating the land to be planted, which stimulates the weeds to germinate. Once the weeds have emerged and are at the seedling stage the crop is drilled, care being taken to minimise disturbance of the soil surface in order to avoid further germination of weed seeds. The emerged seedling weeds are then treated with paraquat two to three days before the crop emerges, which should, in theory, create a weed-free environment for the crop when it emerges.

Since paraquat only affects weeds which have emerged there may be a problem with weeds which germinate after the herbicide has been applied, unless this technique is backed up by the use of appropriate soil-applied herbicides. Where perennial weeds are present presowing or pre-planting treatment with materials like glyphosate, aminotriazole, dalapon and TCA should be considered. It may also be beneficial, where recommended, to apply trifluralin as a pre-sowing/pre-planting treatment for the control of annual weeds.

Effective weed control depends, among other things, on the correct choice of herbicides, which, in turn, demands that the grower should plan in advance which crops are to be grown, and ensure that weeds are identified correctly. As residual herbicides are normally applied to weed-free soil, it is advisable for growers to keep a note of weeds which are associated with particular fields so that, for following crops, appropriate herbicides may be selected.

Incorporation into the soil is essential with a number of herbicides. For example, tri-allate vaporises readily and trifluralin is subject to degradation by the ultra-violet rays of the sun, so that both must be incorporated into the soil. Also, for maximum control of couchgrass TCA should be incorporated to a depth of about 130mm. where it is most likely to encounter the rhizome system of this weed.

The performance of soil-applied herbicides depends, to a great extent, on rainfall and soil type. If there is insufficient rainfall, the chemicals are not transported to the zone where the weeds are germinating, and if the organic matter content of the soil is greater than about 10%, the activity of many herbicides is reduced. One way of overcoming these problems is to incorporate herbicides which do not normally require to be incorporated. This means that rain is not required to carry the material to the weed zone, and there is much less chance of the chemicals being 'bound into' organic soils, and so becoming unavailable.

Although incorporation can enhance the activity of soil-applied herbicides, it must be properly carried out, which means incorporating the chemical evenly across the field at the correct depth. Incorporation at too great a depth means that the majority of annual weeds, which germinate in the top 5cm of the soil, will be missed. On the contrary, if incorporation is not deep enough this can result in crop damage and excessive loss of volatile chemicals.

The additional effort necessary for the incorporation of herbicides, which might deter some growers, may be offset if they can be convinced that this system makes herbicides more efficient and more reliable. Moreover, a sprayer mounted in front of a tractor, with a rotary cultivator at the back, gives good incorporation with very little extra effort.

If the grower is in any doubt whether or not to incorporate a particular herbicide, or if he is unsure about the correct way to carry out the task, he should seek professional advice.

The problem of resistant weeds getting out of hand can be lessened by making use of herbicide mixtures and programmes. Here too planning of crop rotations and accurate weed identification are of paramount importance in order that appropriate combinations of herbicides may be selected.

It should be stressed, again, that growers are not advised to concoct their own mixtures.

Consideration should also be given to the introduction of non-vegetable break crops like cereals, where growth regulator herbicides can be used to help control perennial weeds. Also, annual weeds should not be left to flower and seed after a vegetable crop is harvested: many future weed problems could be avoided if such weeds were sprayed with paraquat before they flowered.

After the chemicals available for any particular situation have been considered, and the most suitable materials chosen, a number of steps can be taken to achieve optimum results.

The spray operator should know exactly what he is supposed to do. This includes reading the manufacturers' instructions and following them rigidly, and checking calculations, calibrations and spray nozzles regularly. Also, the sprayer should be carefully washed out between chemicals.

Correct timing of application is important for most chemicals if they are to be safe to the crop and produce optimum weed control. Fields should, therefore, be examined frequently and thoroughly in order that correct timing can be achieved.

When all these factors have been carefully considered, and if conditions are favourable, then, generally speaking, the earlier spraying takes place the better will be the results.

Some vegetable crops are particularly susceptible to the presence of weeds. Onions and leeks, for instance, are slow to germinate and to become established. Moreover, it takes a long time for these crops to produce a leaf canopy to provide adequate ground cover and, in fact, when they are grown at a wide row spacing they never really achieve this. Weed competition during the early stages of growth of these crops has a devastating effect on yield. Indeed, weeds which are present for seven to eight weeks following 50% crop emergence in spring-sown onions can reduce yield by as much as 60%. For this reason, weed control is important early in the life of these crops. Although it has little or no effect on yield, heavy weed cover later on in the life of the crop can interfere with harvesting. On account of these factors, and because these are high-value crops, it is advisable to keep onions and leeks as free from weeds as possible by making full use of the available herbicides.

In peas, in addition to affecting crop yield, the presence of weeds adds to the expense and difficulties involved in harvesting and vining, and if the level of contamination of the crop by the seed heads of weeds like thistles and poppies is high enough, the crop may not be acceptable.

Carrots are, again, slow to emerge and the foliage they produce does little to suppress weed growth. Moreover, this crop is normally grown on open-textured soils which tend to encourage the rapid emergence and growth of annual weeds.

Although herbicides may not always be as effective as desired, there is little doubt that the growth and yield of vegetable crops would be greatly reduced if chemical weed control methods were not available.

Further Reading

ANON (1980), *ADAS/MAFF Reference Book 95 – Strawberries*, H.M.S.O. Publications, London.

    A useful account of strawberry growing, including weed control.

ANON (1982), *ADAS/MAFF Reference Book 156 – Cane Fruit*, Grower Books, London.

    A helpful guide to cane fruit production, including weed control.

GUNN, E. (1980), 'Economic weed control in top fruit and strawberries', *Proc. 1980 Br. Crop Prot. Conf. – Weeds*, pp. 807–11.

Describes a number of the principles governing weed control in top fruit and strawberries and outlines several herbicide programmes for each situation.

HARDY, F. S. and WATSON, G. D. (1982), *Commercial Vegetable Growing*, Frederick Muller, London.

A comprehensive text covering all aspects of vegetable production, including a useful section on weed control.

MAKEPEACE, R. J. (1976), 'The development of herbicide programmes for field vegetable crops', *Proc. 1976 Br. Crop Prot. Conf. – Weeds*, pp. 915–22.

An account of the principles underlying weed control in vegetables.

ROBINSON, D. W. (1983), 'Herbicide management in apple orchards', *Scientific Horticulture*, *34*, 1983, pp. 12–22.

A concise, well-written review.

11 Amenity grass, ornamental trees and shrubs (including nursery stock) and flower crops

11.1 Amenity grass

The removal of weeds from amenity grass is, perhaps, a somewhat controversial issue. Some would argue that the presence of daisies (*Bellis perennis*) or clover (Trifolium sp.) provides an added attractiveness to the plain green of a lawn, and this may be true. On the other hand, there are those who believe that a lawn should consist only of fine-leaved grasses, and that its only function is to provide a backdrop for the wide array of colours produced by flower and shrub beds. There is no doubt, however, that the presence of plants like stinging nettle (*Urtica dioica*) and spear thistle (*Cirsium vulgare*) on a rugby field, for example, would be acceptable to none but the most ardent participants in this enthralling sport.

Nowadays, therefore, since the cost of herbicides is high, and the cost of labour even higher, there may well be a case for tolerating the presence of certain non-grass plants in turf. However, in situations where weeds are undesirable either because they are thought to be an eyesore, or because they adversely affect the activities for which the turf is used, the careful choice and application of herbicides is a very useful element of any sward-management programme.

With the range of herbicides available for turf providing good control of most common weeds, it is important to discourage the idea that herbicides alone are capable of producing a healthy, luxuriant turf. This is, most definitely, not the case, and herbicides should be integrated into a system of good cultural practice which also includes proper mowing, spiking and aerating, application of fertiliser, watering and the control of pests and diseases.

In the case of amenity grass, weed control may be carried out at two stages. Firstly at the time of site preparation, when every opportunity should be taken to remove problem perennial weeds such as couchgrass, docks and nettles, which could cause a problem later when the new lawn is becoming established. At this time the ground can be cleared by the use of herbicides like glyphosate, dalapon and aminotriazole which normally would be harmful to grass.

Once grass is present herbicide selection becomes much more restricted. On newly sown grass the normal 'lawn' herbicides like 2,4-D, mecoprop and MCPA should not be used. Where only a few annual weeds are present, these are not a problem and will normally disappear when mowing begins. However, if a new lawn is badly infested with annual weeds the contact herbicide ioxynil gives good safe control.

In established turf where weed control is required, this should, so far as possible, be carried out so that damage to the grass is kept to a minimum, and this normally means using selective herbicides.

Turf weeds fall into two groups as follows:

a) Broad-leaved weeds (dicotyledons)
b) Undesirable grasses (monocotyledons).

11.1.1 BROAD-LEAVED WEEDS

In this group selective weed control is achieved, primarily, by virtue of morphological, physiological, or biochemical differences between the broad-leaved weeds and the grasses, which are monocotyledons. For instance, the herbicide 2,4-D is translocated much more slowly in grasses than in broad-leaved weeds, and this is, undoubtedly, an important means of selectivity for weed control in turf.

Selective weed control in established turf is normally based upon the use of the growth-regulator type herbicides like 2,4-D, mecoprop and dicamba, either alone or in mixtures, depending on the weeds present. For instance clover and speedwell (*Veronica* sp.) are resistant to 2,4-D, while creeping buttercup and dandelion are fairly resistant to mecoprop. Although ioxynil is not included in this group, it is often used either by itself or combined with mecoprop for the control of speedwell.

With the exception of ioxynil, the chemicals mentioned previously are all translocated herbicides and enter the plant via the foliage. Thereafter, they move throughout the plant to their sites of action where they exert their herbicidal effects. Because of this, the following points should be borne in mind:

1. The recommended rate of application should be strictly adhered to. If too little material is applied the weeds will not be controlled; if too much, then scorching of the foliage may occur which, in turn, may inhibit translocation of the chemical from the leaf to the rest of the plant.
2. Translocation is greatest when the weeds are actively growing. Therefore, treatment under drought conditions, during periods of dormancy, or even immediately after mowing, may result in poor weed control since insufficient quantities of herbicide will be taken up by the weeds.

3. The process of translocation is not instantaneous and rainfall within eight hours of spraying may reduce the degree of weed control achieved.

Compared with paraquat, the rate at which translocated herbicides act is very slow, and although uptake of growth-regulator herbicides may soon be obvious from the typical leaf malformations which they produce, a period of two to three weeks may be required before the weeds actually die. This period may be even longer if prolonged dry weather follows spraying.

11.1.2 UNDESIRABLE GRASSES

The eradication of weed grasses from lawns and sports fields is considerably more difficult than the removal of broad-leaved weeds. Two of the most common weed grasses are annual meadow-grass (*Poa annua*) and Yorkshire fog (*Holcus lanatus*), a coarse perennial grass. Since these weeds belong to the same group of plants as the desirable grasses (monocotyledons), it is extremely difficult to find herbicides which will remove the weed grasses and leave the others unharmed.

Where weed grasses are present in extensive clumps or patches, it is often more practical to spray with a non-selective herbicide like glyphosate and then to reseed the area.

An added complication with annual meadow-grass is that many greenkeepers and groundsmen consider it to be desirable since it is durable, persistent and spreads readily. This has meant that call for chemicals to control this weed has been by no means unanimous.

Most weed grasses are annuals and as such reproduce from seed, which highlights the possibility of using a herbicide to control the seeds as they germinate. The herbicide ethofumesate may offer considerable assistance in this area. This chemical gives excellent control of annual grasses, including annual meadow-grass, in swards composed mainly of perennial ryegrass (*Lolium perenne*), and if this chemical were to prove equally safe and successful with grass species used in fine turf, it would be an extremely useful addition to the groundsman's armoury. However, at present, ethofumesate is not recommended for use on fine-leaved amenity grasses, and treatment should not be attempted without full consultation with the manufacturer. In the case of Yorkshire fog, asulam appears to give good selective control in perennial ryegrass swards.

In turf, as in other situations, one single herbicide will not normally control all the weeds and so it is essential to spend time identifying the weeds correctly before embarking on any herbicide programme, in order that safe, efficient weed control is achieved.

The presence of moss in turf is generally indicative of inadequate management and, while chemical treatment may be effective in removing the moss intially, unless the underlying causes are remedied and steps taken to promote a healthy, active, competitive turf, it will reappear.

In general, moss thrives in situations where the soil is acid, compacted and badly drained, conditions which produce poor grass growth. Initial preparation of the land to ensure maximum drainage, liming if necessary, the correct use and application of fertilisers, and spiking and raking are all essential for vigorous growth of turf. In addition, mowing is an important factor in moss eradication. Close mowing encourages moss growth at the expense of the grass and should be avoided.

If an effort is made to maintain high standards of turf management, then the chemicals available for the control of moss, like ferrous sulphate and dichlorophen, will do a much better job.

Despite good management practices, it is fairly certain that weeds will, at some time, start to encroach upon grassed areas. In this event, timely and correct application of appropriate herbicides should produce satisfactory weed control.

11.1.3 CONTROL OF VEGETATION

There is no doubt that allowing the vegetation of country parks, roadside verges and embankments to develop completely un-controlled would be welcomed by ecologists and conservationists, and perhaps there is a lot to be said for this point of view. In contrast, however, the public in general and the majority of those responsible for the upkeep of such areas prefer to see them well maintained. Moreover, there are certain situations in which failure to maintain roadside verges and embankments properly could have very serious consequences indeed: for example, where the visibility of motorists is impaired by tall or overhanging vegetation on bends or at road junctions.

In the past, vegetation in these areas was kept under control either by cutting and mowing or by selective grazing. Nowadays, however, the cost of labour makes mowing a less desirable option. In addition, some sites may be either too steep for mowing or unsuitable for the fencing required with grazing.

This means that a replacement method is required and, inevitably, the use of appropriate chemicals must come under consideration. Ideally, any such material should be capable of retarding the growth of vegetation, particularly grasses, and encouraging desirable species. At present, the only chemical approved for suppressing the growth of rough grass is maleic hydrozide.

The initial formulations of this material had a number of disadvantages. There was a real danger of scorch, especially during dry weather. Moreover, during prolonged dry periods the chemical did not produce any response at all. On the other hand, when rain followed spraying the material was readily washed off the treated vegetation, and again the result was little, if any, response.

Formulations of maleic hydrazide now available are much more 'rainfast' and, if applied according to the manufacturers' recommendations, give acceptable results, with the likelihood of scorch occurring being greatly reduced.

Repeated application of this material over a number of years tends to produce a stand of desirable vegetation with dicotyledonous plants as well as grasses. Furthermore, the less tall, more acceptable grasses such as red fescue (*Festuca rubra*), smooth-stalked meadow grass (*Poa pratensis*) and creeping bent (*Agrostis stolonifera*) tend to become dominant at the expense of the undesirable grasses like cocksfoot (*Dactylis glomerata*) and false oat-grass (*Arrhenatherum elatius*). In addition, this chemical gives a degree of control of a number of broad-leaved plants including cow parsley (*Anthrisius sylvestris*), docks (*Rumex* sp.) and hogweed (*Heracleum sphondylium*).

Experience with this material suggests that it is much more effective if applied two to three weeks before mowing. Under these circumstances, mowing can be reduced from once every two weeks to once every ten weeks: there are generally fewer seed heads produced, and the resulting sward is of a much more even height.

The appearance of the sward can be improved still further by the use of a mixture of maleic hydrazide and 2,4-D. The 2,4-D eliminates a variety of broad-leaved weeds which would otherwise produce unsightly flowering shoots.

Another material which is showing promise as a growth retardant in grasses is mefluidide (Marshall, 1981). This material acts more rapidly than maleic hydrazide, is generally reliable, and is particularly effective in suppressing the flowering of fine grasses, which is important in improving the appearance of a sward. Also, mefluidide has the ability to promote the increase in the sward of the more desirable fine-leaved grasses.

It may be helpful, therefore, especially in times of economic cutbacks, to give serious consideration to the use of chemicals as a realistic alternative to mowing for the maintenance of rough grassed areas. This procedure would not inevitably lead to unemployment, as the individuals previously employed in grass cutting could work in other, more productive ways of landscape management.

Furthermore, the use of chemicals for the control of vegetation need not conflict with the views of conservationists and ecologists since these materials, if used properly, are capable of encouraging the development of a thriving mixed plant community.

In order to optimise the degree of weed control obtained without damage to grass, a number of points should be kept in mind, as follows:

Timing of application

Generally, weeds are much more susceptible to herbicides at the seedling stage and, wherever possible, should be treated at this time. Failing this it is essential to treat them before they flower, in order to prevent the production of seeds which would cause a problem at a later date. Spraying should be carried out when temperature and soil moisture levels favour vigorous growth of the grass, so that quick filling in by grasses occurs where the weeds have died out. If bare areas are fairly large they should be re-seeded or re-turfed, since they make ideal seed beds for the germination of weed seeds. If these patches are re-seeded, any soil or compost used either for levelling or mixed with the grass seed should be sterilised, to ensure that no weed seeds are introduced by this means.

As a rule, the efficiency of herbicides increases with increasing temperature. However, if spraying is carried out when the temperature is too high, there is a real danger to the grass. This in turn means that the grass will be less able to fill in any spaces left by the removal of weeds. If, however, there is no alternative to spraying when the temperature is high, consideration should be given to reducing the application rates in order to minimise any harmful effects to the grass.

Dilution rates

Herbicides may be applied directly as supplied at a fixed rate per hectare. Alternatively they are diluted and applied at a fixed rate of concentrate per hectare, in which case the addition of a bulking agent, like water, is purely a means of facilitating the application of a small amount of concentrate over a large area, and the amount of water will vary according to the type of equipment used for application.

Application equipment

The type of equipment used depends both on the formulation of the herbicide and on the area to be treated. For instance, liquids and wettable powders are applied as a spray, using a watering can or knapsack sprayer for small areas. For larger areas, trolley or tractor-mounted sprayers are more appropriate. Granules should be applied evenly either by hand or by means of mechanical spreaders.

It is vitally important that all equipment used for the application of herbicides is properly calibrated. If this is not done, safe and effective weed control will not be possible. The calibration of equipment requires to be adjusted to take account of different situations such as the width of ground to be treated. A knapsack sprayer will need to be re-calibrated each time the spray operator is changed, since walking (spraying) speeds vary, and so the area covered by a constant volume of spray will vary accordingly. Indeed, it is a great and common mistake to assume that the calibration of spraying equipment is a 'once and for all' exercise.

Before any herbicide is applied, the condition of the spraying equipment should be checked for the following points:

1. The spray tank and lines must be clean and not contaminated with the remains of the previous spraying. This could be disastrous if the last chemical used was a total weedkiller like glyphosate, when the job in hand is to apply a selective herbicide to a cricket field. Standard procedure, therefore, should be to wash equipment thoroughly each time it is used.
2. Spray nozzles should be checked each time before use to make certain that they are not cracked or chipped. This will ensure that the correct spray pattern is formed.
3. Pressure controls should be set as recommended on the product leaflet or label.

SPREADERS

1. The hopper should be clean and not contaminated with the material used previously.
2. Check that rollers and grids, normally responsible for metering out the correct dosage rate, are functioning properly.
3. If the application rate is controlled by adjustable levers, ensure that these are securely fixed and cannot alter their setting as a result of vibration when the machine is in operation.

Addition of chemical to the spray tank

Always use clean water, preferably from a mains supply. If it is necessary to use a water course (i.e., a stream or river), the greatest care must be taken to ensure that the water is not contaminated with the chemicals being used. Such contamination could contravene the Control of Pollution Act, 1974.

The best procedure is to half-fill the tank with water and then add the required amount of chemical. The tank is then filled with water to the required level. It is essential that all water and chemical passes through the filter basket in the tank.

When two or more chemicals are being used as a combined spray (tank mix), each concentrate should be added separately, diluted and mixed well before a further concentrate is added.

Wettable powders are best creamed in a bucket with an approximately equal weight of water, in order to ensure that complete wetting of the herbicide takes place. This 'cream' is then added to the water in the spray tank and made up to the required volume.

Empty containers should be washed out thoroughly (at least three times) and the washings added to the tank. Unless empty containers are collected by a waste disposal contractor or by the local authority, they should either be buried or burned. The methods of burying or burning vary with different types of containers, and must be rigidly adhered to. Failure to dispose of pesticide containers in the proper manner, including those used for herbicides, could be an offence under 'The Deposit of Poisonous Waste Act, 1972 and Regulations'.

Information about the disposal of containers and unused chemicals is contained in the MAFF publication, *Guidelines for the disposal of unwanted pesticides and containers on farms and holdings.*

Once the sprayer or spreader has been filled, treatment may begin. Herbicide application should always be carried out methodically and, if necessary, markers used to indicate areas which have been sprayed, and those which have still to be treated. Generally, the wheel marks of the tractor, spraying machine or spreader are sufficient to line up the equipment for each run. In very dry weather, however, when wheel marks are not visible, a small length of heavy chain dragged behind the equipment will leave a suitable guiding line.

Where high quality turf is being sprayed, it may be advisable to apply two half-strength doses at right angles to one another. If this method is used, the area still receives the correct dose of herbicide, but the split application greatly reduces the risk of uneven spraying and damage to the turf.

The danger of herbicide drift must always be borne in mind by the spray operator, and much can be done to overcome this problem by using low pressure anti-drift nozzles, and applying chemicals at a high volume of water per hectare. Nevertheless, changes in the weather should be noted and spraying stopped immediately conditions become unsuitable.

Since controlled droplet application (CDA) of herbicides was developed, a number of new techniques for the application of herbicides have appeared, including the rope-wick, or wiper applicator. The underlying principle of this method of application is that non-selective herbicides can be applied selectively, and tall weeds can be controlled in a low-growing crop by means of a concentrated formulation of a translocated herbicide like glyphosate.

The rope-wick applicator comprises a pipe reservoir-boom containing the herbicide, mounted on a frame. A number of nylon rope wicks are strung along the boom and threaded through compression joints into the boom, which allows herbicide to permeate the wick both by capillary action and gravitational flow. Internal seals prevent herbicide dripping from the boom.

As the tractor is driven over the area to be treated, the saturated ropes wipe herbicide directly on to tall weeds growing above the crop. With this equipment translocated herbicides such as glyphosate may be used, and no re-formulation is required. In the UK a rope-wick applicator is marketed by Hectaspan as the 'Weed Wiper'.

The recommended herbicide for use with the Weed Wiper is

glyphosate, applied in line with normal product recommendations. This is useful, among other things, for the control of the majority of grassland weeds.

11.2 Ornamental trees and shrubs

11.2.1 NURSERY STOCK

It is generally accepted that weeds compete with crop plants for a number of factors including light, water and nutrients, but the actual extent by which weeds reduce the growth and value of crops, including trees and shrubs, is seldom fully appreciated.

Results of trials using bare root liners (small plants with no soil on their roots), albeit with a small number of species, highlight the devastating effect that weed growth can have on lined-out nursery stock plants (Davison, 1976). This work revealed that failure to control weeds in the year of planting, even though the plots were kept weed-free in the second year, reduced the value of Norway maple (*Acer platanoides*), Lawson's cypress (*Chamaecyparis lawsoniana*), mock orange (*Philadelphus virginalis*) and *Potentilla fruticosa* cv. Primrose Beauty, by between 30 and 50 per cent when the plants were lifted after two seasons. The effect on laurel (*Prunus laurocerasus* cv. Zabeliana) was even more striking. In this case no plants of a marketable size were obtained two years after planting, from plots which were weedy in the planting year. Furthermore, those plants which remained, and were grown on, were worth only about 15 per cent of those from weed-free plots.

The reduction in value of *Acer*, *Potentilla* and *Philadelphus* grown on weedy plots resulted from the fact that the plants were smaller: the presence of weeds did not, in fact, lead to plant death in these species. By contrast, in Lawson's cypress and *Prunus*, weed growth, in addition to reducing the size of the plants, resulted in losses of 30 per cent and 60 per cent respectively.

Admittedly, these results refer only to a limited number of species grown in one particular area, and in two exceptionally dry summers. However, the effects of the presence of weeds on the growth and ultimate value of the test species, together with similar results in other crops, are so striking that they undoubtedly provide food for thought for any conscientious nurseryman.

In general, the species and cultivars of ornamental trees and shrubs used in amenity plantings are the same as those grown in commercial nurseries. One important difference, however, is that whereas in nursery stock production large areas are planted with the same species or cultivars, in the amenity situation there are usually a number of different species and cultivars planted together. Furthermore, the economics of the two situations are different since the livelihood of the nursery stock producer depends on his producing more attractive, better quality material than his competitors. These two factors mean that a different strategy must be adopted when it comes to weed control.

In the UK many ornamental trees and shrubs, as well as all forestry-type conifiers, are raised from seed and lined out for one or two years after which they are marketed. These plants are small and, as such, are particularly susceptible to weed competition. For this reason, in commercial nursery stock production, it may be advisable, although expensive, to consider using soil sterilant materials like dazomet and methyl bromide prior to planting high-value stock. These chemicals, if applied correctly, give excellent control of the vast majority of pests, pathogens, weed seeds and portions of vegetative plant material with which they come into contact. This ensures that young plants can get off to a very good start.

There are, however, a number of disadvantages with these chemicals. The cost is fairly high at around £1,000 per ha for dazomet, which does not include the cost of polythene to seal the surface of the soil once the sterilant has been incorporated. While this is not essential, and rolling may be adequate under certain circumstances, it is generally more efficient. The application of methyl bromide requires to be carried out by a contractor, and, in many cases, a contractor is also employed in the case of dazomet treatment. This, again, adds to the cost of these materials.

A major problem with these chemicals is that they have no long-lasting effect. Four weeks after dazomet application (longer in cold soils) the polythene is removed and the ground thoroughly cultivated to remove all traces of the chemical (a cress germination test is advisable in order to ensure that all the chemical has disappeared). Any weed seeds which then enter the treated area will not be controlled.

At present, there are no recommendations on herbicides which can be used after seeds of ornamental trees and shrubs have germinated, and it is essential, therefore, to prepare a clean seed bed before sowing. If the cost of using a soil sterilant is prohibitive, the land should be fallowed for at least a season, when any weeds should be dealt with by the use of appropriate materials such as paraquat or glyphosate. At this time it may also be helpful to use a relatively non-persistent residual herbicide to control germinating annual weeds. This gives the perennials a chance to produce adequate foliage for treatment with glyphosate or aminotriazole. If grasses are the main problem, dalapon can be used.

It may, however, prove very difficult to stimulate perennial weeds to produce sufficient foliage during a one-year fallow, in which case chemicals like glyphosate may give less than satisfactory control of these weeds.

Whenever it is feasible, therefore, the use of a soil sterilant material such as dazomet on nursery stock seed beds might prove beneficial in the long term even though this method is initially expensive. Such materials if used properly produce excellent results, and weeds such as groundsel, the seeds of which come in on the wind and germinate, can be pulled by hand.

In the case of transplanted trees and shrubs, the position with regard to herbicides is much more encouraging, but even here only chlorthal-dimethyl with propachlor, and diphenamid are approved for application on ornamental trees and shrubs immediately after planting. In addition, there is an ACAS provisional approval for trifluralin (formulated as Treflan) as a pre-planting treatment. Therefore, because there is only a very limited choice of herbicides recommended for use at the establishment stage of nursery stock plants, every effort should be made to eradicate weeds, especially perennials, before the stock is planted.

In the past, mechanical methods of weed control, including hoeing, were widely used in nursery stock. These methods are still in widespread use, but at present-day costs hand labour, if it can be found, is now a much less attractive proposition. Furthermore, hoeing and other mechanical techniques are not only less efficient than herbicides, but can seriously affect the growth and development of trees and shrubs, as a result of the damage they cause to surface roots (Fig. 11.1).

A promising alternative to mechanical and chemical weed control methods is the use of black polythene as a mulch. Admittedly, there are a number of problems associated with this method. There is a fairly hefty financial outlay initially for the polythene (around £800 per hectare), and there are undoubted practical problems with laying polythene over a large area.

Against this, the black polythene is long-lasting, and will remain usable until it is degraded, or becomes so badly torn that it no longer provides satisfactory weed control. In addition to providing effective weed control, a mulch of black polythene may increase the growth of plants compared with those kept weed-free by other methods. This, in turn, may lead to increased profits by providing a better-sized, more saleable product, as reported by Davison, 1976 and Davison and Bailey, 1980. These studies revealed that the value of crops mulched with black polythene increased by £300/ac. in the case of *Philadelphus*, and by £1,500/ac. in the case of *Prunus*, and that many mulched plants of *Potentilla* and *Philadelphus* reached a saleable size after one season's growth.

It may be, therefore, that black polythene, despite its cost, could provide an effective and, in the long term, a profitable alternative for the grower who is prepared to experiment with this material, and show some initiative to overcome the problems of laying this material over a large area.

Fig. 11.1: *A blackcurrant bush grown in herbicide-treated non-cultivated soil (bottom) will be more vigorous than a bush grown in cultivated soil (top) because the surface roots are able to exploit the fertile surface soil layers (courtesy D. W. Robinson).*

There are also a number of alternatives to black polythene which may be used for mulching, including straw, grass mowings, sawdust, peat and pulverised bark. Each of these materials has associated disadvantages. For instance, straw is rather unsightly, peat is expensive and sawdust tends to cause the soil to become compacted under wet conditions.

However, all these materials provide undoubted benefits when used as a mulch. For example, by providing a barrier at the soil surface, they play an important role in conserving moisture, in the upper layers of the soil, which would normally be lost by evaporation.

In addition, as long as enough material is provided organic mulches will, in general, give good control of annual weeds. Provided black polythene is not torn or does not contain too many perforations (to allow rain to penetrate) it will give excellent control of weeds: clear polythene, however, is totally ineffective for this task. Moreover, a mulch of polythene can be effective in raising the temperature of the soil with the result that nitrification is improved. Crop growth is also improved, which may, in turn, lead to an earlier crop and increased yield.

11.2.2 CONTAINER-GROWN PLANTS

Although there is an increasing tendency towards the production of nursery stock plants in containers, and a variety of valuable plants are now grown in this manner, this sector of crop production is not large enough to merit special attention by the agrochemical industry.

The major sources of weeds of container plants are the compost in the containers, the sand on which the containers stand, and the land around the nursery.

While certain weeds may be introduced in the compost, the ones which cause most trouble are those which produce large numbers of wind-borne seeds such as hairy bitter-cress (*Cardamine hirsuta*), groundsel (*Senecio vulgaris*) and annual meadow-grass (*Poa annua*). Seeds of these and other weeds may be blown on to the sand on the standing ground. Treatment with paraquat as a directed spray normally provides safe and effective control, but glyphosate is not advised, as it may be taken up by the roots of the container plants and so cause severe damage.

The two herbicides most commonly used in controlling weeds of container plants are simazine and chloroxuron. These are applied as overall sprays, every five or six weeks in the case of chloroxuron and every nine weeks or so for simazine. Where chloroxuron is used, it is normal practice to wash the foliage of the container plants immediately after treatment, in order to avoid damage. However, care should be taken to avoid using too much water as this may lead to herbicide being leached to the plant roots, producing harmful effects.

In addition to controlling problem weeds like bitter-cress and willow herb (*Epilobium* sp.), chloroxuron also gives good control of mosses and liverworts. Annual meadow-grass, however, is only moderately susceptible.

Although simazine does not require to be washed off the plants, it does not control mosses and liverworts, and it may cause damage to a number of species of nursery stock plants.

Other possible herbicide treatments in this situation include low rates of propyzamide where grasses are a problem. In addition, diphenamid may be a useful alternative in containers, but it does not control bitter-cress, one of the most important weeds of container-grown stock. Also, it requires adequate moisture and temperature in order to be really effective.

A recent introduction for the control of annual grass and broad-leaved weeds in field and container-grown nursery stock is napropamide, either alone or as a mixture with simazine.

Napropamide on its own can be used on a wide range of container and field-grown nursery stock and controls a large number of annual weed species including cleavers, annual meadow-grass, groundsel and black bindweed. If it is combined with simazine, control of shepherd's purse and bitter-cress is also achieved.

Where the mixture is used, certain simazine-sensitive species may be affected and the mixture should not be used on these plants. These include certain varieties of *Betula, Deutzia, Euonymus, Forsythia, Fraxinus, Ligustrum, Lonicera, Prunus, Senecio, Spiraea, Syringa, Viburnum*, and *Weigela*.

These materials should be applied to weed-free soil (emerged weeds are not controlled) and may be broken down by the action of sunlight; application during summer months is therefore not recommended.

Ideally, application should be made between November and February, when winter rainfall should carry the herbicide into the weed zone. However, if the soil can be irrigated, the herbicide may be applied through to March and April.

There are a number of problems associated with the use of the knapsack sprayer for the application of herbicides to container-grown plants. These include damage to the crop as a result of uneven spray cover, and high cost due to the time required to carry out the operation.

A promising alternative system has been developed by the ADAS mechanisation department, however, which consists of a pedestrian-carry spray boom operated by a mobile or tractor-mounted spray reservoir, with a pump at one end of the bed. For herbicide application the spray boom is carried by two people, one at each side of the bed to be sprayed, who transport the boom along the length of the bed.

With the use of commercially available 'Rain Drop' nozzles which give a large droplet size, better penetration of herbicide spray may be obtained than with the knapsack sprayer. The number of nozzles may be varied to suit each particular situation, but the maximum number recommended is fifteen.

Other advantages of this system are that it is cheap and easy to construct and it reduces dramatically the time spent on weed control on one acre of containers over the year (18 man hours) compared with hand weeding (1,884 man hours) and knapsack spraying (144 hours).

The land around the standing ground may present a potential weed problem. For instance, if this ground does not belong to the grower and weeds are left to develop uncontrolled, it will become a source of vast numbers of weed seeds, which may enter the standing ground.

If, however, this land does belong to the grower and is not under cultivation, care must be taken to ensure that weeds in this area are controlled before they flower and produce seed. Appropriate herbicides for total weed control on non-cropped land will be discussed later.

11.2.3 AMENITY PLANTINGS

Moving on to the amenity situation, there appears to be a certain reluctance on the part of parks departments and other similar organisations to adopt a weed management system based on herbicides. In a survey of Scottish Local Authorities (Woods, 1980 and Murray and Woods, 1981), paraquat was found to be virtually the only herbicide used for clearing the land prior to planting with ornamentals (Fig. 11.2). Furthermore, while a slightly greater range of materials was used by local authorities in established trees and shrubs (Fig. 11.3), paraquat was still, by far, the most commonly used herbicide. Although paraquat is a very useful herbicide for the control of annual weeds, its value against perennial weeds is somewhat limited.

If only landscape managers could be convinced of the value of herbicides as a low-cost alternative in the maintenance of trees and shrubs, and be persuaded to design plantings in parks and gardens which base their weed control primarily on the use of herbicides! All too frequently, it appears that such plantings are designed without any consideration being given to the use of herbicides, and trees and shrubs which do not tolerate the same chemicals are, in many instances, planted side by side.

There may be a number of reasons why landscape managers are reluctant to turn to chemical methods of weed control, such as a basic lack of appreciation of what herbicides can achieve if they are properly used. Moreover, there appears to be a commonly-held view that while herbicides are well suited for use in commercial horticulture, they are fraught with difficulties when it comes to amenity situations. A further problem may be the apparent discrepancy in the standards of training

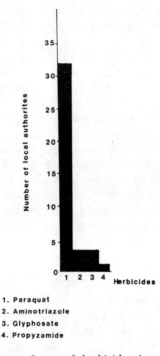

1. Paraquat
2. Aminotriazole
3. Glyphosate
4. Propyzamide

Fig. 11.2: *Frequency of use of herbicides in pre-planting ground clearance (out of 40 local authorities who used herbicides in this situation) (Murray and Woods, 1981).*

received by individuals involved in applying herbicides. For instance, in the survey undertaken by Woods in 1980, although the majority of authorities indicated that they used only trained staff to apply herbicides, the description of the type of formal training that spray operators received was, in general, vague. Descriptions of training such as 'on the job' training, verbal instructions and practical demonstrations were common, and only a few authorities indicated that a comprehensive training scheme was offered to employees who used herbicides.

At first sight there would appear to be a number of major problems associated with the use of herbicides in amenity areas. However, on closer examination the position is a little more hopeful.

For instance, annuals and herbaceous plants, widely used in amenity plantings, are sensitive to most of the common herbicides, but if woody ornamentals were used instead this problem would be largely overcome, since these plants are tolerant of simazine and other soil-applied herbicides.

When designing amenity plantings based on the use of herbicides,

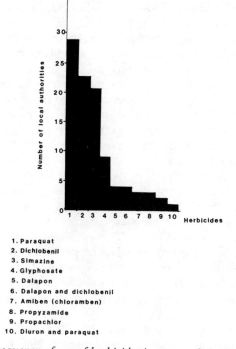

1. Paraquat
2. Dichlobenil
3. Simazine
4. Glyphosate
5. Dalapon
6. Dalapon and dichlobenil
7. Amiben (chloramben)
8. Propyzamide
9. Propachlor
10. Diuron and paraquat

Fig. 11.3: *Frequency of use of herbicides in trees and shrubs (out of 40 local authorities who used herbicides in this situation) (Murray and Woods, 1981).*

two main options are available regarding systems of planting.

Firstly, once the herbicide programme has been decided, trees and shrubs which tolerate the various herbicides to be used can be selected and planted in appropriate groupings. Alternatively, since there is little lateral movement with modern soil-applied herbicides, a wide range of trees and shrubs may be planted together, and those which are sensitive to any particular chemical in the programme can be surrounded by a mulch of black polythene. This will provide effective weed control until the plants become established, at which time they are less susceptible to herbicide damage. Once the polythene is in place, the rest of the area may be sprayed with appropriate chemicals.

In many instances the tolerance of ornamental trees and shrubs to herbicides is due to depth protection, which means that the plants' roots are below the level at which the herbicide is to be found in the soil. Consequently, even if an area has been treated with a soil-applied herbicide, it is still possible to introduce additional trees and shrubs (Fig. 11.4). First, the top 5cm of soil (A) is removed and placed to one side. Next, a hole large enough to accommodate the plant is dug and this soil (B) placed on the other side. After the plant has been inserted, the herbicide-free soil (B) is placed around the roots. Finally, the surface soil containing herbicide (A) is replaced on the surface, where it should continue to provide effective weed control.

There is evidence (Robinson, 1975) that treatment of a large number of woody ornamental plants (over two hundred genera and species) in a garden, with two applications of simazine per year over a number of years, causes very little damage to the plants. Although the foliage of wild strawberry (*Fragaria vesca*) was severely damaged, and some herbaceous species like phlox showed symptoms of simazine damage each year, none of the vast number of woody species treated showed any adverse effects. Indeed, genera such as *Syringa*, *Prunus*, *Senecio* and *Spiraea*, normally believed to be particularly sensitive to simazine, grew well and did not show any symptoms of herbicide damage on the foliage. Moreover, contrary to expectation, there was no damage to *Erica* and a number of non-woody plants including *Bergenia*, *Sedum*, and *Dianthus*. Nor, apparently, was there any effect on flower size, flower number, or the vigour of bulbs such as narcissus, tulip, crocus and snowdrop, after six years of repeated applications of simazine. These satisfactory results are still being obtained after fourteen years.

While the indiscriminate use of herbicides over large areas of ornamental plantings is not advocated, these results would suggest that if landscape managers were to experiment with herbicides, even on a small scale, the scope for using chemical methods of weed control in amenity horticulture could be greatly increased.

Should a landscape manager wish to adopt a policy of weed control based on the use of herbicides, how may this best be achieved?

Each amenity area requires a tailor-made herbicide programme. The choice of herbicides depends on the ornamental plants to be used, and on the weeds which, judging from past experience, are likely to occur. A major factor in the success of any such programme is to choose herbicides which will complement each other's spectrum of weed control. This system is not as difficult to follow in practice as it may seem at first sight. The important point is that a number of herbicides are used each year. For example, a soil-acting herbicide like simazine, which is applied to control germinating weed seeds, should be supplemented with appropriate translocated materials such as 2,4-D or contact herbicides like paraquat. If a combination of herbicides plus judicious hand-weeding to prevent weeds from flowering and seeding is used, there is much less chance of weeds getting out of control.

A few guide-lines for the maintenance of tree and shrub plantings, based mainly on soil-applied herbicides, may be helpful.

As far as possible, always plant into clean ground, making full use of

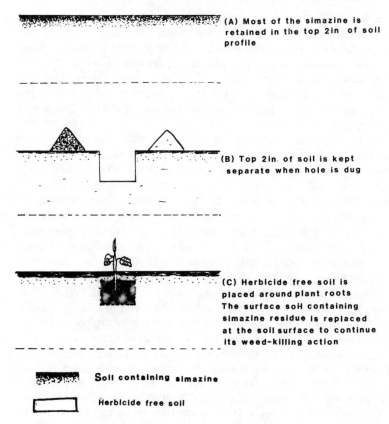

(A) Most of the simazine is retained in the top 2in of soil profile

(B) Top 2in. of soil is kept separate when hole is dug

(C) Herbicide free soil is placed around plant roots The surface soil containing simazine residue is replaced at the soil surface to continue its weed-killing action

Soil containing simazine

Herbicide free soil

Fig. 11.4: *Method of planting into herbicide-treated soil (courtesy D. W. Robinson).*

herbicides like glyphosate, aminotriazole and dalapon. After planting, a soil-acting herbicide such as simazine should be applied as an overall treatment. Any weeds which appear after this can be spot-treated with appropriate chemicals such as paraquat, glyphosate, MCPA or 2,4-D. Although there is a risk of damage from growth-regulator type herbicides such as MCPA and 2,4-D, if these materials are applied as a low pressure spray in calm weather, they can be used effectively around many ornamental trees and shrubs. The risk of damage by these chemicals may also be reduced if they are applied when perennial weeds are still small, or in the autumn when, although many ornamentals have stopped growing, weeds will still be active.

While, as mentioned previously, successful weed control in amenity plantings generally depends on the eradication of perennial weeds before planting, the new selective grass-killers alloxydim-sodium and fluazifop butyl now make possible the control of couch and other perennial grasses in established plantings.

It is also desirable to eliminate hoeing, for two main reasons. When herbicides are applied to the soil they form a 'seal' which prevents germinating weeds from emerging. Hoeing breaks this 'seal' and stimulates the germination of weed seeds. Furthermore, hoeing tends to damage the surface roots of ornamental plants, and so reduces their growth and development.

In the past, soil which was hoed was generally considered to be 'healthy'. However, continual use of herbicides over a number of years produces a more compacted surface which tends to become covered with moss, and this leads to the commonly held belief that such herbicide-treated soil is 'unhealthy' and, therefore, bad for plants. In reality, however, moss-covered soil, far from being unsuitable for the growth of trees and shrubs, may well be beneficial through preventing erosion and protecting the soil from rain damage. In addition, because the use of herbicides eliminates hoeing, plant roots are not being disturbed and so are able to absorb more nutrients from the soil.

This philosophy demands a change in attitude, not only in the general public but also in many local authorities, towards the use of herbicides and soil appearance, which will, it is to be hoped, eventually come about, although this may take a number of years.

The use of herbicides around ornamental trees and shrubs could be much more effective if local authorities would adopt a policy of using properly trained 'weed control squads', who, because they used equipment on a regular basis, would be much more efficient and, by the same token, less likely to cause any damage to non-target plants.

There is no reason, therefore, why herbicides should not provide an effective, safe, and reasonably cheap alternative to conventional methods of weed control in amenity plantings, especially if such areas are designed with the use of herbicides as an integral part of management practice from the outset.

11.3 Weed control in flower crops

Chemical weed control in flower crops presents a number of problems. For instance, many different species and varieties of flowers are grown, and the area under cropping and also the market for flower crops are quite small.

This means that no chemical company or other body will devote much time or allocate substantial resources towards testing different herbicides in flower crops.

As a result, much of the information available concerning herbicides for use in flower crops has arisen from trials on other crops, which indicate that particular chemicals may be suitable for flowers. Moreover, an essential source of information on this subject is the growers themselves, a number of whom are prepared to experiment with materials on a small scale, and then pass on successful findings to fellow-growers, advisers and chemical manufacturers.

A number of herbicides are approved for use in flower crops, such as dazomet for soil sterilisation and glyphosate and TCA for couch control pre-planting. For general weed control, materials like chlorpropham mixtures; chlorthal-dimethyl and chlorthal-dimethyl + propachlor; diphenamid; propachlor and paraquat (+ mixtures) may be used. In addition, trifluralin has provisional clearance under ACAS for use in flower crops.

These, however, are only a few of the herbicides which are approved for flower crops, and a full list of approved materials can be found in the ACAS booklet.

The dose at which soil-applied herbicides are used selectively in horticultural situations does not normally control established weed seedlings. This is due, in part, to the fact that the roots of established weeds are usually too deep in the soil to come into contact with the herbicide. If soil-applied herbicides are used at the minimum recommended dose for selective weed control, there should therefore, under normal circumstances, be little risk of damage to established flower crops.

Growers should not be afraid to experiment with soil-applied herbicides in different flower crops. That being said, however, it is in no way suggested that an entire crop be sprayed with a chemical the effects of which are unknown. A small area only should be treated in the first instance and this should be undertaken, whenever possible, in consultation with a government adviser or a trials officer from the company which manufactures the material.

Further Reading

DAVISON, J. G. (1983), 'Effective weed control in amenity plantings', *Scientific Horticulture*, 1983, *34*, pp. 28–34.

This article provides a full account of the topic covering various aspects such as methods of weed control, costs, strategies and herbicide programmes.

MCCAVISH, W.and INSLEY, H. (1981), 'Give Trees a Chance', *GC and HTJ*, April 24, 1981, pp. 26–8.

This article looks at the safe use of herbicides around broad-leaved amenity trees and its particularly useful in that it contains a section on calculation of herbicide concentration.

ROBINSON, D. W. (1983), 'The use of herbicides in amenity horticulture', *Scientific Horticulture*, *34*, 1983, pp. 23–7.

Gives an overall picture of some of the problems associated with the use of herbicides in amenity horticulture including landscape fashions, factors limiting progress and weed control programmes.

ROBINSON, D. W. (1975), 'Herbicides in the landscape garden' (Parts I, II and III), *Garden*, *100* (11), pp. 554–9; *100* (12), pp. 600–606; *101* (1), pp. 35–41.

This series of articles gives a basic comprehensive account of herbicides including selectivity and mode of action. In addition it describes the use and effects of herbicides in a large garden, and discusses the possible hazardous effects of herbicides on the soil and micro-organisms.

ROBINSON, D. W. (1983), 'Weed control in nursery stock and amenity plantings: In FLETCHER, W. W. (ed.), *Recent Advances in Weed Research*, C.A.B. pp. 119–226.

A comprehensive review of weed control in field and container-growth nursery stock and in amenity plantings.

STEPHENS, R. J. and THODAY, P. R. (eds) (1980), *Proc. Conf. on Weed Control in Amenity Plantings*, University of Bath.

A first-class review of weed control in amenity plantings covering a wide range of topics including the weed problem; perennial weeds; strategies and programmes; mode of action; application of herbicides; total weed control and present and future role of herbicides in amenity land management.

12 Non-crop situations

12.1 Weed control around glasshouses

Vapour drift (see Chapter 4) is not caused by growth-regulator herbicides alone. Under appropriate conditions (warm dry weather) vapour from dichlobenil may drift into glasshouses and cause damage to susceptible crops. It is recommended, therefore, that dichlobenil is not used to control weeds around glasshouses.

The control of weeds around glasshouses is, however, important since they may encroach, either by seed or by vegetative means, into the glasshouse. Also, weeds in this situation provide an ideal environment for pests and diseases which, under suitable conditions, will attack the plants in the glasshouse.

However, there is a danger in using a herbicide in this situation either from drift, or because the material may move laterally in the soil and so enter the glasshouse soil. Therefore herbicides which are considered suitable for use around glasshouses must possess particular characteristics, including lack of persistence in the soil and low volatility. They should also be effective against a broad spectrum of weeds, and be of low toxicity to the operator, who may be applying the chemical in a fairly confined situation.

Perhaps the herbicide which fits this description best is glyphosate although it is not, at present, recommended for this purpose. Even with glyphosate, however, application should be made, if possible, at a time when there are no plants in the glasshouse, Failing this, the vents must be tightly closed. If the glasshouse is known to leak, then an alternative method to the use of herbicides must be adopted.

If the situation is suitable for the use of herbicides, they should be applied in a form in which the number of drift-prone droplets is kept to a minumum.

There are a number of ways of achieving this, such as the use of high volume, low pressure spray nozzles. An alternative is to make use of a controlled droplet application (CDA) spraying technique. This system offers the possibility of using only 20–40 litres of liquid carrier per hectare and the equipment is light-weight, which allows wet land to be sprayed.

CDA depends upon the production of herbicide droplets the size of which is controlled within a narrow range. This uniformity in drop size is achieved by feeding the herbicide solution on to a serrated-edged, fast-rotating disc which breaks up the solution into a mist-like curtain of drops. The drop size is dependent upon the flow rate of the herbicide and the speed of the rotating disc. After the spray droplets have left the rotating disc, they are carried to the target area by the wind and by gravity.

Conventional spraying using hydraulic pressure nozzles produces a wide range of droplet sizes. These vary from very small droplets of about $70\mu m$ ($1\mu m = {}^1/1000mm$) which give good crop coverage but are very susceptible to both spray drift and vapour drift, to very large droplets of around $400\mu m$ which tend to bounce off the target. On the other hand, with CDA droplet size is governed, with the vast majority of droplets being around $250\mu m$ in diameter. Droplets of this size are more resistant to evaporation and seem to provide a reasonable crop cover with little drifting, except in quite windy conditions.

One of the most commonly used CDA sprayers is the Micron 'Herbi' which uses a single 8cm diameter, unshrouded, electrically-driven disc.

A further interesting method of application which may be worth considering is to apply glyphosate in the form of a thin gel. Advantages of this method include greater retention on plants which are not easily wetted, and it is less likely to be lost from the treated plants if rain follows spraying.

12.2 Weed control on non-cropped land (total control of vegetation)

There are a number of situations where the eradication of all vegetation, including weeds, is required. These include pathways and driveways, railway tracks, industrial installations and certain sports surfaces. In these areas the presence of weeds is not only unsightly but may be extremely dangerous. For instance, heavy weed growth on land surrounding petrol storage tanks could present a very real fire risk during prolonged dry weather.

In these situations, the requirement is for all existing vegetation to be removed, after which the land must be kept free from weeds.

While hoeing or burning may achieve temporary control of weeds in such situations, the most effective method for clearing the land initially, and keeping it weed-free in the long term, is the use of herbicides.

For clearing land of established weeds, translocated herbicides like glyphosate, aminotriazole, dalapon and the growth-regulator type materials are invaluable. However, these chemicals have no effect on weeds that have still to emerge from the soil. It is necessary, therefore, to use chemicals which remain in the soil over a long period of time, in order to control germinating weeds.

Such herbicides include picloram, bromacil, dichlobenil, simazine, chlorthiamid and sodium chlorate. The concentrations at which these herbicides are used for total weed control demand that the greatest care must be taken with their application. This is especially true in the case of bromacil and sodium chlorate, which are extremely prone to lateral movement, or to migration down a slope, both of which can harm any crop or ornamental plants which lie in their path.

It is now common practice to use combined sprays of translocated and soil-applied herbicides, which makes the whole procedure more efficient.

Generally, this method of using herbicide mixtures provides a high standard of weed control, and should any weeds emerge, they can be dealt with by spot-treatment with appropriate materials.

There are also many occasions, such as clearing ground before planting a crop or sowing a lawn, when it is important to remove existing vegetation. In such situations it is neither necessary nor desirable for the herbicides to remain in the soil for any length of time.

Herbicides such as glyphosate or paraquat, if annual weeds are the main problem, are useful in such situations, since these chemicals are virutally non-persistent.

12.3 The control of aquatic weeds

Even though weeds are found in virtually all expanses of fresh water, they require to be controlled only when they present problems, such as blocking drainage ditches or the water intakes for industrial plants. Water weeds may also hinder other activities for which the water is used, such as fishing or boating.

It is important to determine whether these plants are a nuisance before steps are taken to eradicate them, since they may have a vital role in the maintenance of the water body by providing, among other things, oxygen as a result of photosynthesis, and shelter for a number of aquatic animals. In addition, water weeds may play an essential part in the prevention of erosion and slipping of the beds and banks of streams and other water courses.

For convenience, aquatic weeds are grouped in the following categories:

Emergent weeds

This group includes plants such as reeds and rushes whose leaves and stems are held above the water surface. It also includes broad-leaved plants like water plantain (*Alisma plantago-aquatica*) and arrowhead (*Sagittaria sagittifolia*). A number of the broad-leaved plants found on riverbanks are also included, for example hairy willow-herb (*Epilobium hirsutum*).

Floating weeds

This group comprises plants which have at least some of their leaves floating on the surface of the water. While the plants in this group are generally rooted on the bottom like yellow water-lily (*Nuphar lutea*), some, such as duckweed (*Lemna* sp.) are free-floating.

Submerged weeds

The plants in this group are generally totally submerged except at flowering when, in most cases, the flowering shoots appear above the surface of the water. Again, most of these plants, (water crowfoot (*Ranunculus aquatilis*) for instance) are attached to the bottom, but one or two like ivy-leaved duckweed (*Lemna trisulca*) float freely under the water surface.

Algae

A number of different types of algae may, on occasion, present a weed problem. The most common are the single-celled microscopic algae which, under favourable conditions, multiply very quickly, producing what are known as 'algal blooms'. In addition, green filamentous algae may grow together in a tangled, troublesome mat ('blanket weed') and in this form these primitive plants may be responsible for impeding water flow in drainage ditches, and for blocking water intakes in industrial plants.

As with land plants, the aquatic weeds which are the greatest nuisance are the perennials. Although a few emergent weeds may produce vast numbers of seeds, for example over 200,000 from one flower-head of the great reedmace (*Typha latifolia*), the majority of perennial aquatic weeds reproduce vegetatively from their stems or from rhizomes, and dispersal of vegetative fragments from one body of water to another takes place via the water itself, or by other agencies such as birds, animals and man.

Like that of terrestrial plants, the rate of growth of aquatic plants is related to the availability of nutrients, carbon dioxide, light and appropriate temperature. In conditions in which all of these factors are favourable, the growth of aquatic weeds will be at an optimum and, consequently, the need for weed control will be greatest.

AMOUNT OF WEED CONTROL REQUIRED

Before deciding on any programme for the control of aquatic weeds, it is important to ascertain how much vegetation does, in fact, need to be removed. For instance, only in situations where the risk of flooding is very high is the removal of all vegetation necessary.

With this in mind, the following classes are designated for the control of aquatic weeds.

Complete eradication

In urban areas which are susceptible to flooding it may be necessary to remove all the vegetation, and then to keep the area weed-free on a permanent basis.

Controlled growth

This involves removing excess growth from an area without significantly reducing the number of plants present, and is most often undertaken for the improvement of surface drainage and in areas where fishing is important.

Occasional control

This is practised, in general, where the management of wildlife is of paramount importance, and it involves the periodic thinning out of vegetation.

METHODS OF WEED CONTROL

In any given situation, a number of factors influence the method of weed control which would be most appropriate. These include the degree of control required, the characteristics of the water and the operations for which it is used, the availability and cost of labour, and any practical difficulties which may be encountered in reaching the weeds.

Once the decision has been made as to the degree of weed control necessary, a number of techniques are available.

Cutting and raking

This system is employed in situations where the complete eradication of weeds in the long term is not required. The weeds are cut either by hand or by using one or more of a number of boat or tractor-mounted machines, after which the cut material is removed by means of a rake and deposited on the bank. It is essential to remove the cut vegetation from the water since as it rots and is broken down, it utilises oxygen which is dissolved in the water. This can lead to deoxygenation, which is dangerous to fish and other aquatic animals. Moreover, if cut vegetation is not removed, this could present other problems such as the blockage of ditches, channels, sluices and water intakes. Also, care must be taken to cause as little disturbance to wildlife as is consistent with obtaining satisfactory control, and cutting at nesting time, for instance, should be avoided.

Although this method of weed control may be popular with anglers and conservationists, if offers only temporary control, and there are many instances where the weeds may not be accessible.

Other methods include dredging and burning but these techniques also have associated problems. Dredging is a slow, costly business and it does not give good control of the roots and rhizomes of perennial weeds. In addition, the mud and other dredged material deposited on adjacent farmland can be a source of weeds which will require to be controlled at a later date.

Where only temporary vegetation control is necessary burning may be considered, but this technique has to be repeated frequently, is ineffective against perennial weeds, and is most useful as a follow-up treatment to the use of herbicides.

Biological control

This technique involves using one organism to control another which is normally a pest. In the aquatic situation, the most common method employed for the biological control of weeds in the UK is selective grazing and treading by stock (mainly sheep) on river banks and verges. However, in areas where the banks are particularly steep, this can lead to erosion.

Another possibility for biological control of aquatic weeds is the use of certain species of fish which will selectively graze on the weed plants, and this method may prove valuable in the future.

Chemical weed control methods

The use of herbicides to control aquatic weeds has a number of advantages over other methods. For instance, by careful choice and application of herbicides it is possible to produce either a temporary reduction in the growth of vegetation or, alternatively, to obtain complete eradication of water weeds over long periods. Furthermore, the use of herbicides in the aquatic situation is much less labour-intensive than mechanical methods and, nowadays, may be just as economical.

Generally, two types of herbicides are used in aquatic weed control: those which are sprayed on to the foliage of emergent or floating weeds, and those which are applied directly to the water and are taken up from the mud. The latter type of material is used mainly in the control of submerged weeds, against which foliage-applied herbicides would be ineffective. Also, this type of chemical is less selective than foliage-applied materials.

The addition of a chemical to a watercourse may cause a number of problems and a few points should be kept in mind before embarking on the use of herbicides for the control of aquatic weeds, including:

1) the types of weeds to be controlled and their correct identification
2) the characteristics of the water (e.g. slow or fast-flowing)
3) whether adjacent susceptible crops might be at risk from spray drift
4) is the water used for recreational purposes? is it drinking water for humans or livestock? or is it used for irrigation?
5) will there be any danger to fish and other aquatic life? For instance, the herbicides themselves or some component of the formulation may be toxic to fish. On the other hand, the harmful effects may be indirect, as a result of deoxygenation of the water caused by decaying vegetation.

The timing of herbicide treatment is crucial and, as a rule, chemicals which are added to the water itself are usually applied in the spring, while materials which act via the foliage are normally applied in the late summer or autumn. This being said, however, there is less danger to fish as a result of deoxygenation if weeds are treated before their growth becomes excessive. Also, whenever possible any area of water should be treated in sections, as this gives fish a chance to migrate to an untreated area if the amount of dissolved oxygen falls to an unacceptable level.

Before herbicides are used in or near water, the organisation responsible for the area of water to be treated must be notified. In England and Wales this is normally the appropriate Water Authority, and in Scotland it is usually the River Purification Boards. In Northern Ireland, the appropriate body to inform in such situations is the Water Service for Northern Ireland. The addresses of these organisations are contained in MAFF Booklet 2078, *Guidelines for the use of herbicides in or near watercourses and lakes*.

These authorities are able to provide relevant information regarding the various uses of the water to be treated. It is then the responsibility of the herbicide user to ensure that there are no adverse effects on any of these uses.

Furthermore, before a herbicide is applied to any body of water it should be ascertained whether any site of special scientific interest, or any nature reserve, would be in danger from this action.

For this reason, it is essential to inform the Nature Conservancy Council or whoever else has responsibility for such designated areas, in order that their precise situation may be determined.

The use of herbicides for the control of aquatic weeds, as for all other purposes, is subject to a number of items of legislation concerning safety to the user, the public, wildlife, the control of environmental pollution, and the maintenance of water quality. Information on these and other aspects of the chemical control of aquatic weeds is found in MAFF Publication 2078, mentioned previously.

This situation, perhaps more than any other, demands that careful thought be given to the method of weed control to be adopted. Then, if the use of herbicides is considered to be the most appropriate technique, only chemicals which have been cleared under PSPS for the control of aquatic weeds should be used, and the manufacturers' recommendations followed to the letter.

Chemicals which are, at present, approved for use in or near water include diquat and terbutryne for the control of algae; maleic hydrazide and mixtures for vegetation control on riverbanks and verges. In addition, dalapon, dichlobenil, glyphosate, chlorthiamid, diquat and terbutryne are available for use against emergent, floating and sumberged weed species.

A number of different spraying methods may be employed in the application of herbicides in or near water. These include knapsack sprayers, tractor-operated sprayers, boat-mounted sprayers and aerial application.

Information on these methods and other factors involved in the use of aquatic herbicides, such as dose rate, weeds controlled, time of application, toxicity, and any special requirements relating to these materials, is available from the manufacturer.

Further Reading

ANON(1973). *Control of Aquatic Weeds*, MAFF/ADAS Reference Book 194, HMSO, London.

> This booklet provides information on the principles and practice of aquatic weed control including the biology of aquatic weeds, methods of weed control, and legislation relating to the use of herbicides in aquatic situations.

DAVIES, A. (1982), 'All Under Control'. *GC and HTJ*, March 19, 1982.

> This article covers concisely and in an easily understood manner the calibration of knapsack sprayers and the best way of applying herbicides in horticultural situations.

STEPHENS, R. J. and THODAY, P. R. (eds), (1980), *Proc. Conf. on Weed Control in Amenity Plantings*, University of Bath.

> Included in these Proceedings are articles on 'Total weed control on hard landscape using controlled droplet application' and 'Total weed control on hard lanscape areas'.

Conclusion

A number of pointers as to the general direction in which weed control is currently moving have been indicated in the text. Attention has also been drawn to certain areas in which progress is particularly rapid at the present time. These include, for example, application technology, where we can expect continued developments in reduced-volume spraying, both with conventional and rotary-atomiser-based equipment. Improved droplet control by means of electric charging is an especially interesting field and could lead to significant improvements in efficiency (reduced wastage) and in biological performance, if problems of crop penetration and placement can be overcome. Reductions in dosage-rates may be less reliably obtained, until underlying principles have been more fully explored, although recent work in sugar-beet offers at least one way forward.

In chemicals, the most important development perhaps centres on the continuing introduction of the new graminicides for the control of weed grasses and volunteer cereals in broad-leaved crops. The selectivity and flexibility of these materials make them a very valuable addition to the weed control armoury. The same is true of another interesting group, the sulphonyl-ureas, whose activity at very low dosage-rates is a major feature. Finally, while winter cereals will probably continue to receive special attention, at the other end of the scale the introduction of metazochlor and similar chemicals in swedes may help to revive the fortunes of this potentially valuable crop.

Longer-term developments in weed science and the use of herbicides are more problematical. However, a few suggestions may be offered, based on current trends. Agriculture in its present, highly productive form came about largely on a basis of low-cost oil. It is estimated that the farming industry accounts for about 4% of primary fuel consumption in the UK and that this figure rises to around 13% when food processing and distribution are included. Although herbicides account for only a small proportion of this energy, the matter is especially pertinent to weed science since the main raw materials involved are, of course, hydrocarbons. It seems likely, therefore, that in time the increasing cost and scarcity of fossil fuels will demand the greater involvement of materials such as the natural plant oils, which are already coming into use, for example, as alternative carriers to water in some CDA applications.

It is anticipated that even at the farm level all operations will be carefully monitored and that less energy-demanding methods, such as direct-drilling, will become more widespread. Although this may require increased use of herbicides, the total energy requirement will be less, due to the more favourable energy input/output ratio resulting from the substantial yield increases associated with chemical weed control. Improved spray techniques and light-weight (l.g.p.) vehicles are examples of innovations in which continuing progress may be expected.

On the other hand, however, minimal cultivations must also be numbered among those farming practices with harmful side-effects, in this case the increased occurrence of crop plants as weeds. The varied nature of this particular problem, including both cereals and broad-leaved crops, demands a multi-disciplinary solution. Plant breeders, for example, will have to produce cereal cultivars which do not shed their grains and which do not become dormant in response to environmental conditions. Also, more precise information will be required from weed scientists and agronomists regarding post-harvest treatments, in order that plant propagules from previous crops can be more effectively destroyed. More efficient harvesting machinery, more specific herbicides, and equipment which allows non-selective herbicides to be used selectively, will all have a part to play.

One aspect of weed biology which has received considerable attention world-wide, i.e. herbicide resistance, has shown little sign of developing in Britain, despite thirty-five years of increasingly intensive herbicide usage. Unlike some insects and disease-causing organisms, few British weeds have more than one generation per season which means that it would take a relatively long time for a weed's resistance to a given herbicide to show up. It is possible, however, that this problem will increase and continued vigilance will be required in monitoring the situation.

While genetically resistant weed populations are therefore still very uncommon in this country, practices such as minimal cultivations in cereals and bed-systems in vegetables have made us much more dependent on the use of herbicides. This has led to the rapid increase of formerly unimportant species such as field pansy, for example, following the repeated use of urea-based herbicides in winter wheat. More attention may have to be paid in these situations, therefore, to the role of cultural methods, especially crop rotations, in addition to the use of herbicides in mixtures and sequences.

Also, after the regular application of herbicide to a particular area, weeds may not appear sufficiently numerous to warrant further treatment. In this case it may be tempting for farmers or growers not to spray, which is likely to result in a renewed build-up of weed populations. There has been much discussion of spraying thresholds, in the major crops especially, but while some have been established with reasonable accuracy (e.g. annual grass weeds in cereals), the effects of broad-leaved weeds, as outlined in the text, do not usually lend themselves to this kind of approach and much work remains to be done in this area.

To date, no major problems have arisen as a result of the synergistic effects of tank mixes of herbicides or the cumulative effects of a series of herbicide treatments – both practices which seem certain to increase in popularity. This is due, perhaps, more to luck than good judgement, and there is an urgent need for information, both from industry and from the advisory sector, on herbicide compatibility in tank mixes and sequences. As the age of 'prescription' weed control draws near, when increasingly specific recommendations will be made for weed problems, there will be a need for specialist weed advisers to provide weed control programmes to cater for individual problems.

There are several areas where technological advances will be instrumental in dealing with future weed science problems. As our understanding of the biochemistry and physiology of plant growth and development improves, it may be possible to design herbicides to act specifically on marginal biochemical differences between crop and weed. There will be a greater use of computers to identify chemical structures and to design herbicides to act on specific biological processes in plants. As computers become generally available, advisory methods are already changing. The personal computer of a farmer or grower will be able to access data from a central bank, for specific weed problems in specific crops and cultivars, covering factors such as soil type, stage of growth of crop and weed, application rates, results of experimental work and cost.

There will be an expansion in the use of herbicide antidotes or 'safeners', which will help to overcome problems of selectivity between crops and related weeds. Ideally, herbicides should persist long enough to exert their desired effects and thereafter be degraded to non-toxic residues. Future developments will enable season-long control of many weeds by means of controlled release formulations, which, although more costly than other types, require lower doses and fewer applications.

It is likely that there will be an increase in the use of application equipment, such as the rope wick applicator, which can be used to apply non-selective herbicides selectively and economically. At the same time, emphasis will switch to the plant breeder, to produce cultivars of crop plants resistant to the commonly-used herbicides. This has already been achieved to a limited extent with cultivars of perennial ryegrass which are resistant to paraquat, and certain strawberry cultivars which tolerate simazine. As the techniques of genetic engineering improve, it is possible that herbicide-resistant crop plants will be developed by the transfer of the gene or genes controlling resistance.

Although it is possible to achieve almost total weed control in some high-value crops such as raspberries, apples and nursery stock, there seems to be a reluctance to aim for weed-free crops more widely. Complete eradication of weeds would present major problems, such as erosion on sloping sites, reduction of pollinating insects essential to fruit set in a number of crops, and also removal of food sources of parasites and predators of crop pests. However, as advances are made in herbicide technology and biological control there may well be some moves in this direction. Indeed, at present, organic mulches are used successfully to prevent erosion in fruit plantations on moderately sloping sites as part of an overall herbicide management system.

Inherent in any policy of 'weed free' crops, however, must be a determination on the part of farmers and growers to integrate a conservation aspect into their management programmes. The formation of local Farming, Forestry and Wildlife Advisory Groups (FFWAGS) in the UK offers an opportunity for liaison between all those with a vested interest in this area.

As society advances and high technology industries lead to an ever-increasing amount of leisure time, a greater proportion of the population will have the opportunity to develop new interests such as ornamental and vegetable gardening. This could greatly increase sales of herbicides to the amateur and would, of necessity, require new developments in packaging, marketing and education.

The use of herbicides in amenity situations could also be increased, but this would require a change in attitudes towards the use of chemicals. There is a general concern and lack of understanding about the effects of herbicides on human and animal health, desired plants and soil structure. Too much emphasis has been placed on the adverse effects of herbicides on the environment and too little on the part they can play in enhancing it. For instance, with very few exceptions, herbicides have a remarkable safety record and this needs to be widely publicised, especially in view of the fact that as public interest in the environment increases, so the use of pesticides will come under even greater scrutiny.

Injury to plants caused by repeated cultivation (including hoeing), although not so obvious as herbicide damage, is still widespread. Cultivation not only causes root injury but removes surface roots, which means that plants are unable to exploit the fertile soil surface region. A change will be required in the commonly-held view that cultivated soil is 'healthy' and therefore good for ornamental plants, and that a crusty, moss-covered, herbicide-treated soil is detrimental to plant growth. However, in order to cater for public taste many land managers still cultivate weed-free herbicide-treated soil. The problem of the appearance of such soils can be overcome by using low-growing groundcover plants and organic mulches such as peat and pulverised bark in combination with herbicides.

If, in future, advances in herbicide technology create problems such as environmental pollution and resistant weeds, it is not unreasonable to assume that the technology will also develop to overcome them. Given a more positive approach to the use of herbicides and a willingness to accept the small degree of risk involved, herbicides can bring enormous rewards, both in crop production and in improved amenity.

Further Reading

BODE, L. E and BUTLER. B. J. (1983), 'New techniques and equipment for ground application of herbicides in the U.S.A.', *Proc. 10th Int. Congr. of Plant Protection*, Brighton, Nov. 1983, pp. 478–85.

A recent general account, equally applicable to the UK.

FINNEY, J. R. (1979), 'Future weed problems and their control', *Ann. appl. Biol. 91 (1)*, pp. 144–60

A concise account of the major aspects, including the use of herbicide antidotes, plant breeding and integrated control systems.

ROBINSON, D. W. (1978). 'The challenge of the next generation of weed problems', *The Sixth Bawden Lecture, Proc. 1978 Br. Crop Prot. Conf. – Weeds, pp. 800–21.*

A comprehensive review of potential problems facing all those concerned with crop protection, particularly weed control. A number of possible solutions are offered for consideration along with discussion of the prospects for a 'weed-free environment' and weed control in a 'post-industrial society'.

ROBINSON, D. W. (1980), 'Present and future role of herbicides in amenity land management', *Proc. Conf. Weed Control in Amenity Plantings, University of Bath, 1980,* compiled by R. J. Stephens and P. R. Thoday.

This article sets out the present situation with regard to the effects of herbicides in amenity plantings, including long-term effects, and suggests a number of ways in which such herbicides will be of benefit in in the future.

STEPHENS, R. J. (1982), *Theory and Practice of Weed Control*, The Macmillan Press Ltd., London and Basingstoke, pp. 194–204.

A useful account of future developments in weed control including herbicide application technology and changes in crop agronomy.

Appendices

Appendix 1: Weed seedling guide

WEED SEEDLING IDENTIFICATION

It is important to be able to identify weed seedlings as accurately and as early as possible, in order that the correct herbicide may be chosen; by and large, the younger the seedling the better the control obtained.

Weed seedlings comprise three parts, all of which can assist in identification:

(i) The *cotyledons* are in all cases the first part of the seedling to appear above the soil and are always in *opposite pairs*; they therefore lie *below* the true leaves. Like the true leaves, the cotyledons vary in size and shape, although to a lesser extent; a few are also hairy. In some, especially the rosette-forming species, the cotyledons may soon be lost.

(ii) The *true leaves* resemble the leaves of the mature plant, although they are normally smaller and may differ in shape (e.g. poppies). They normally arise singly – the *alternate* arrangement – (e.g. red-shank, charlock), or in *opposite pairs*, often at right-angles to one another (e.g. nettles, speedwells). In some species, however, the basal (radical) leaves arise successively round the stem, forming a circle of leaves or '*leaf-rosette*' (e.g. shepherd's purse, mayweeds). Occasionally the leaves are *whorled*, arising in groups of four or more at each node (e.g. cleavers).

The true leaves vary mainly in *size* and especially *shape*, including the existence of *teeth* or *lobes* (leaves with a smooth outline are said to be *entire*). The presence of *hairs* (and occasionally the *type* of hair) is another important feature and in a few cases *colour* is also helpful. The typical features may develop only gradually, so that the first leaf or pair of leaves differs from the later ones (e.g. mayweeds, fat hen); in other species all the true leaves are alike from the start (e.g. nettles, polygonums).

(iii) The *hypocotyl* is the stem of the seedling *below* the cotyledons. It varies in *length*, being prominent in some cases (e.g. fumitory, corn gromwell) and almost non-existent in others, so that the seedling appears to rest on the soil surface (e.g. mayweeds, shepherd's purse). In a few species the colour is distinctive (e.g. redshank). Fig. A.1 (below) illustrates these main features:

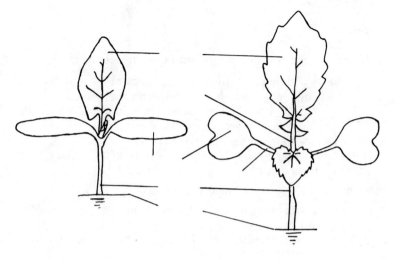

Fig. A.1: *Main features of weed seedlings.*

LAYOUT OF THE GUIDE

The Guide includes the main annual weeds of cultivated land, together with a number of others of local or increasing importance. A few longer-lived species particularly liable to develop from seed have also been included, but not those which regenerate primarily from vegetative fragments (e.g. creeping thistle, field bindweed, white clover). Related species have been noted, where appropriate, and in all some 60 weed seedlings may be identified using this Guide.

The seedlings have been arranged in seven broad groups, based mainly on cotyledon type. Within each group similar seedlings have been placed together and attention drawn by pointer to their most distinctive features. A completely 'tidy' arrangement is clearly impossible, however, and on occasion overall similarity has dictated the placing. All the seedlings are 1.5 times actual size. The symbol ± means 'more or less'.

Unless otherwise indicated, the view of the seedlings is as seen from above. In a few cases (e.g. fat hen, orache) leaves have been 'removed' in order to show the cotyledons. Certain particularly distinctive features have been underlined and some others, not evident from the drawings, included in parentheses. In general, plants may be taken as being non-hairy (glabrous), unless stated to the contrary.

The authors are particularly grateful to Mr R. J. Chancellor and to Blackwell Scientific Publications for permission to use these drawings.

SYNOPSIS

A. Mainly rosette-forming plants, with short hypocotyls.

  A.1 Plants with the first true leaves ± linear or lobed, later leaves finely divided; small, sessile cotyledons (mayweeds).

  A.2 Plants with less divided leaves with broad terminal lobes (includes the cut-leaved forms of shepherd's purse).

  A.3 Plants with generally entire true leaves.

  A.4 Plants with entire, hairy, densely-crowded true leaves; rosette less developed.

B. Plants with elongate or linear cotyledons.

  B.1 Plants with elongate cotyledons and alternate, ± pointed true leaves.

  B.2 As above, but plants 'mealy', especially around the growing point.

  B.3 Plants with short, linear cotyledons.

  B.4 Plants with long, strap-shaped cotyledons and toothed or divided true leaves.

C. Plants with mainly broad-ovate cotyledons.

  C.1 Plants with blunt, stalked cotyledons and ± pointed true leaves.

  C.2 Plants with ± rectangular, stalked cotyledons and indented or wavy-edged true leaves.

  C.3 Plants with sessile cotyledons and toothed true leaves.

D. Plants with large, prominent cotyledons.

E. Plants with rounded cotyledons and opposite true leaves.

F. Plants with deeply-toothed, ± hairy true leaves.

G. Plants with pointed cotyledons and true leaves.

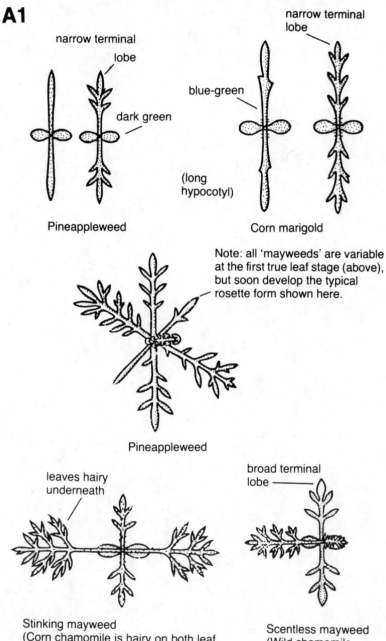

A1

narrow terminal lobe

narrow terminal lobe

blue-green

dark green

(long hypocotyl)

Pineappleweed

Corn marigold

Note: all 'mayweeds' are variable at the first true leaf stage (above), but soon develop the typical rosette form shown here.

Pineappleweed

leaves hairy underneath

broad terminal lobe

Stinking mayweed
(Corn chamomile is hairy on both leaf surfaces)

Scentless mayweed
(Wild chamomile is similar)

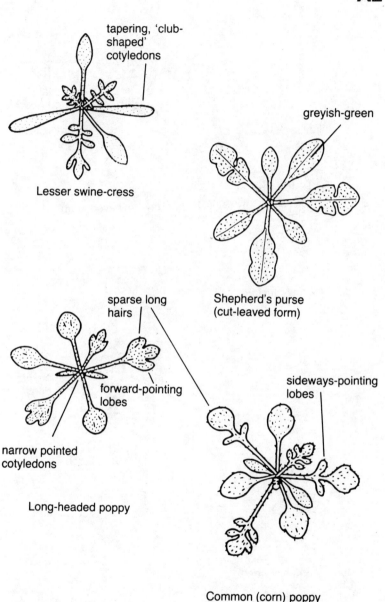

tapering, 'club-shaped' cotyledons

Lesser swine-cress

greyish-green

Shepherd's purse
(cut-leaved form)

sparse long hairs

forward-pointing lobes

narrow pointed cotyledons

Long-headed poppy

sideways-pointing lobes

Common (corn) poppy
(There is also a form with more divided leaves).

A3

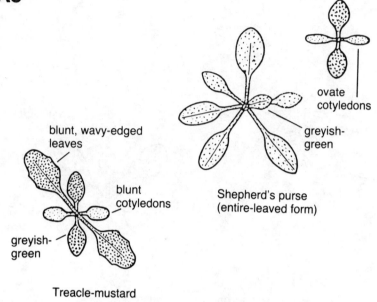

blunt, wavy-edged leaves

blunt cotyledons

greyish-green

Treacle-mustard

ovate cotyledons

greyish-green

Shepherd's purse (entire-leaved form)

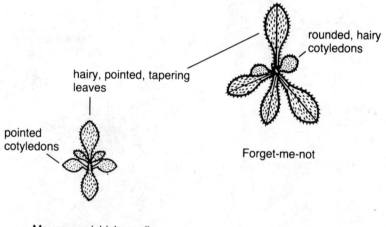

hairy, pointed, tapering leaves

rounded, hairy cotyledons

pointed cotyledons

Forget-me-not

Mouse ear (chickweed) (oblique view)

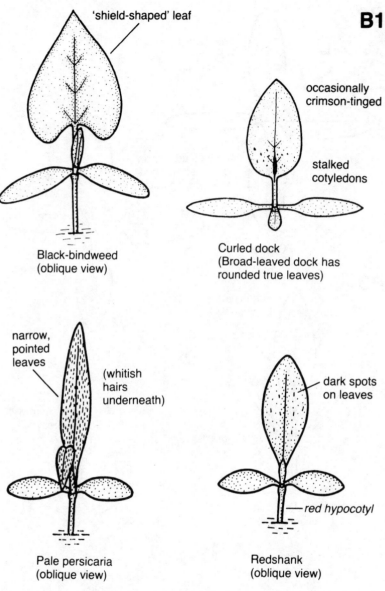

'shield-shaped' leaf

occasionally crimson-tinged

stalked cotyledons

Black-bindweed
(oblique view)

Curled dock
(Broad-leaved dock has
rounded true leaves)

narrow, pointed leaves

(whitish hairs underneath)

dark spots on leaves

red hypocotyl

Pale persicaria
(oblique view)

Redshank
(oblique view)

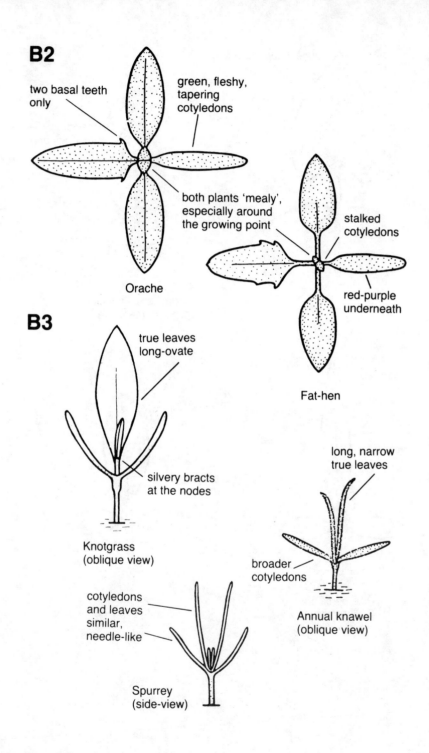

B2

two basal teeth only

green, fleshy, tapering cotyledons

both plants 'mealy', especially around the growing point

Orache

stalked cotyledons

red-purple underneath

Fat-hen

B3

true leaves long-ovate

silvery bracts at the nodes

Knotgrass (oblique view)

long, narrow true leaves

broader cotyledons

Annual knawel (oblique view)

cotyledons and leaves similar, needle-like

Spurrey (side-view)

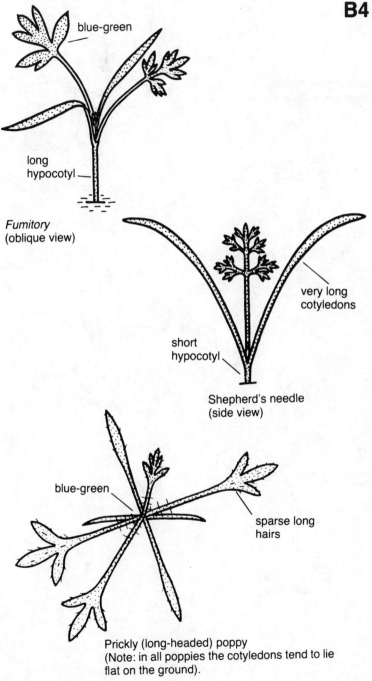

blue-green

long
hypocotyl

Fumitory
(oblique view)

very long
cotyledons

short
hypocotyl

Shepherd's needle
(side view)

blue-green

sparse long
hairs

Prickly (long-headed) poppy
(Note: in all poppies the cotyledons tend to lie
flat on the ground).

C1

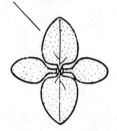

rounded, sessile leaves, often purple-tinged

Many-seeded goosefoot

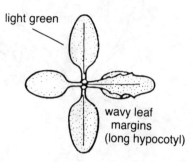

light green

wavy leaf margins (long hypocotyl)

Field penny-cress

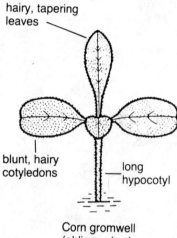

hairy, tapering leaves

blunt, hairy cotyledons

long hypocotyl

Corn gromwell (oblique view)

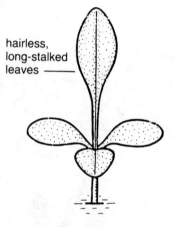

hairless, long-stalked leaves

Hoary pepperwort (oblique view)

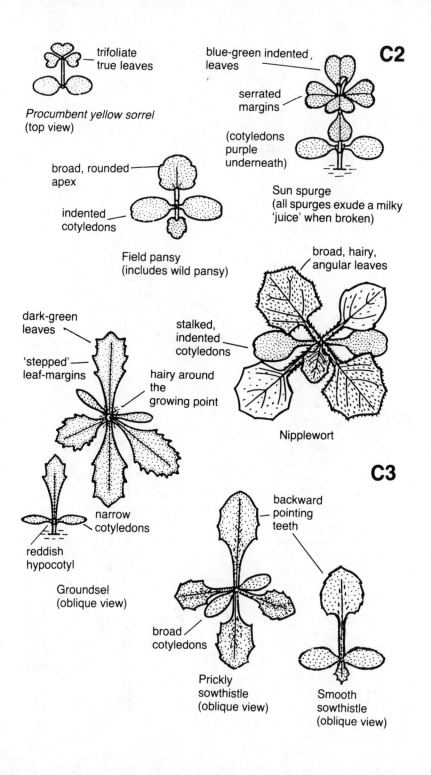

trifoliate true leaves

Procumbent yellow sorrel (top view)

blue-green indented leaves

serrated margins

(cotyledons purple underneath)

C2

Sun spurge
(all spurges exude a milky 'juice' when broken)

broad, rounded apex

indented cotyledons

Field pansy
(includes wild pansy)

broad, hairy, angular leaves

dark-green leaves

stalked, indented cotyledons

'stepped' leaf-margins

hairy around the growing point

Nipplewort

C3

narrow cotyledons

reddish hypocotyl

Groundsel
(oblique view)

backward pointing teeth

broad cotyledons

Prickly sowthistle
(oblique view)

Smooth sowthistle
(oblique view)

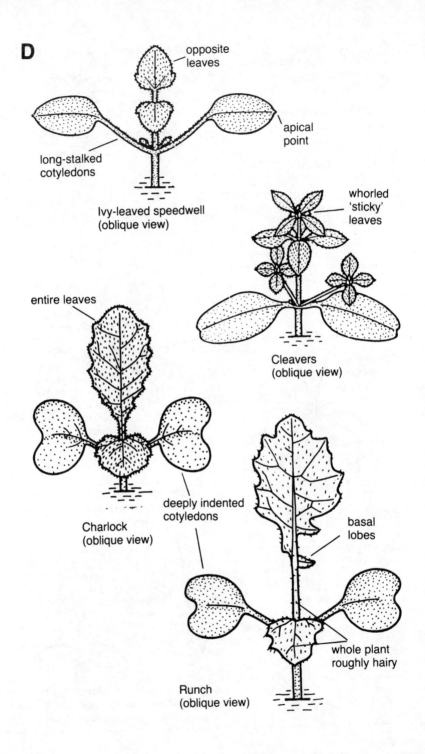

D

opposite leaves

apical point

long-stalked cotyledons

Ivy-leaved speedwell
(oblique view)

whorled 'sticky' leaves

Cleavers
(oblique view)

entire leaves

deeply indented cotyledons

Charlock
(oblique view)

basal lobes

whole plant roughly hairy

Runch
(oblique view)

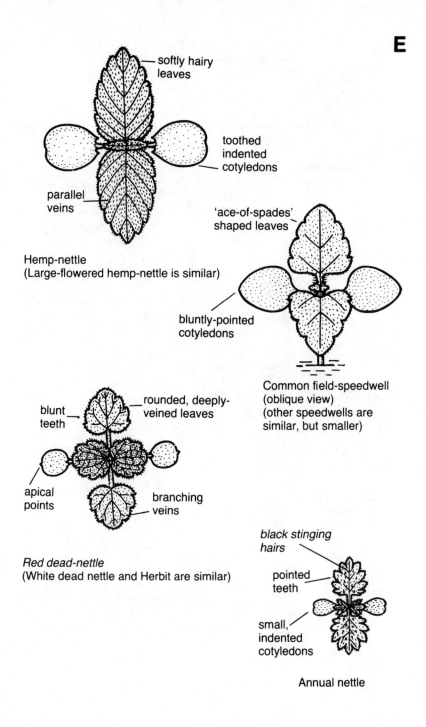

E

softly hairy leaves

toothed indented cotyledons

parallel veins

Hemp-nettle
(Large-flowered hemp-nettle is similar)

'ace-of-spades' shaped leaves

bluntly-pointed cotyledons

Common field-speedwell
(oblique view)
(other speedwells are similar, but smaller)

blunt teeth

rounded, deeply-veined leaves

apical points

branching veins

Red dead-nettle
(White dead nettle and Herbit are similar)

black stinging hairs

pointed teeth

small, indented cotyledons

Annual nettle

F

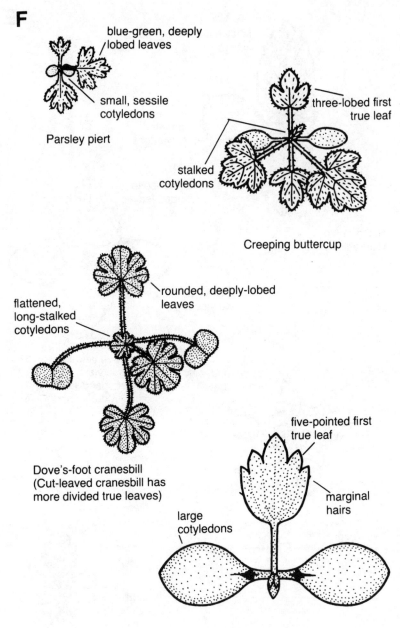

blue-green, deeply lobed leaves

small, sessile cotyledons

Parsley piert

three-lobed first true leaf

stalked cotyledons

Creeping buttercup

rounded, deeply-lobed leaves

flattened, long-stalked cotyledons

Dove's-foot cranesbill
(Cut-leaved cranesbill has more divided true leaves)

five-pointed first true leaf

marginal hairs

large cotyledons

Corn buttercup

G

dark green, triangular leaves
and cotyledons

Scarlet pimpernel

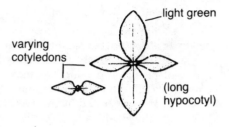

light green

varying
cotyledons

(long
hypocotyl)

Yellow toadflax
(top view)

light green

stalked
cotyledons

long hypocotyl

Chickweed
(oblique view)

leaves hairy,
often purple
below

hairy
cotyledons

Black nightshade

(cotyledons
flat to the
ground)

rough, hairy
blue-green
leaves

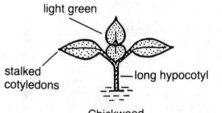

bluntly pointed
hairy
cotyledons

Bugloss
(top view)

Appendix 2: Regulations and Legislation relating to the use of Herbicides

In the UK the supply of pesticides (including herbicides) is controlled by two schemes formally agreed between the Government and the pesticide industry (but see below).

The Pesticides Safety Precautions Scheme (PSPS) deals with the safety of pesticide products as they affect man, livestock, domestic animals, wildlife and the environment.

The PSPS is operated by the Ministry of Agriculture, Fisheries and Food, and the Health and Safety Executive, acting on the advice of the independent Advisory Committee on Pesticides (ACP).

The remit of the ACP is to take stock of the dangers which may arise from the use of the following categories of products:

— Pesticide chemicals (including herbicides)
— Veterinary chemicals not directly administered to animals
— Any other potentially toxic chemical specifically referred to the Committee by Ministers.

The Committee presently comprises scientists who are independent of Government and commercial organisations and members drawn from a number of Government Departments.

A company wishing to sell a new pesticide product, or launch a new use for a product which has already been cleared, must notify the PSPS and provide enough information for potential risks to be assessed and appropriate safety recommendations made, concerning the safe use of the product. These data must include information on the properties of the product, persistence of the product in crop plants, in the soil and in water, and the breakdown products of the chemical and its mode of action. In addition, information on the toxicity of the product to mammals and its effects on wildlife including birds, bees and fish must also be presented.

If the data are acceptable, the ACP will recommend safety precautions and other conditions of use. Acting on these recommendations, the PSPS will indicate suitable label recommendations and will grant clearance for use of products in successive stages from small-scale trials clearance, limited clearance (for large trials, involving limited sale of the product), to full commercial clearance for full-scale marketing.

The second scheme is the Agricultural Chemicals Approval Scheme (ACAS) which deals with the efficiency of pesticide products, and only products which have been fully cleared under PSPS can be considered for approval under ACAS.

It is not necessary for manufacturers to submit pesticide products for clearance under ACAS and, indeed, there are a number of very good chemicals on the market which have not passed through ACAS.

To obtain approval for a product a company must supply information on product composition, shelf life and compatibility. Details of field trials which corroborate claims made for the efficiency of the product must also be provided, and ACAS officers inspect a representative sample of trials in progress.

The main aim of ACAS is to provide farmers, growers and advisers with impartial information about chemicals for use in agriculture and horticulture, in order that they may select safe and efficient products for a particular situation. This being the case, considerable attention is given to label instructions, and before approval is finally given there must be agreement between ACAS and the manufacturers concerning the claims and directions for use which appear on the product label.

Once agreement has been reached, ACAS will grant approval for specific uses of a product under UK conditions sometimes in specific regions. The product label may then bear the 'Crowned A' symbol, and will be listed in MAFF Reference Book 380, *Approved Products for Farmers and Growers*, which is revised annually.

Provisional Acceptance

Provisional clearance may be given under ACAS for the minor use of products which have previously been cleared under the scheme as being safe and efficient in certain specified situations. Provisional Acceptance may be granted in cases where the supporting evidence, although not sufficient to obtain full approval under ACAS, suggests that the product will be safe and efficient in the crop concerned. Minor uses of products are indicated on the label as 'Uses Provisionally Accepted under the Agricultural Chemicals Approval Scheme'.

BRITISH AGROCHEMICAL SUPPLY INDUSTRY SCHEME LTD (BASIS)

BASIS was set up in March 1978 as a registration scheme for distributors of crop protection products selling to farmers, growers and foresters. The Scheme is administered by a Registration Board and has the support of the British Agrochemicals Association, whose members have agreed that they will not supply distributors with crop protection products unless they are registered. The Scheme requires distributors to sell only those products cleared under PSPS; it sets minimum standards for the storage of pesticides, and it has established minimum qualifications for those whose job it is to advise farmers on the use of pesticides.

UNITED KINGDOM LEGISLATION

The Pesticides Safety Precautions Scheme itself has no direct legal status. A number of items of legislation do exist, however, which relate to the safe use of pesticide products, and which are taken into consideration before clearance is given under the Scheme. These include the following;

Health and Safety at Work etc. Act, 1974.
This Act imposes general obligations as follows:

a) on employers, to ensure so far as is reasonably practicable the health, safety and welfare at work of their employees.
b) on the self-employed and employees to take reasonable care of their own health and the safety of others. The Act also requires suppliers to ensure that substances are safe and without risk to health when properly used.

The Health and Safety (Agriculture) (Poisonous Substances) Regulations, 1975.
These set operator protection requirements (age limits, protective clothing, hours of work, training and supervision, etc.) concerning the use of the more toxic products.

The Poisons Rules, 1972.
These control the storage and sale of pesticide products which are scheduled as poisons.

The Deposit of Poisonous Waste Act, 1972 and Regulations.
Under this Act it is an offence:

a) to deposit on land waste of any kind which is poisonous, noxious or polluting so as to create an environmental hazard, except in accordance with a licence.
b) to remove from any premises or deposit on land any waste without notifying the local authorities – unless the waste is of a description exempted from this requirement by Regulations made under the Act.

Regulations under this Act release farmers from a notification procedure if they dispose of pesticides or substances used for the treatment of animals in any safe manner on agricultural land. An empty cleaned container would similarly be exempt from the requirement to notify.

Rivers (Prevention of Pollution) Act 1951 and 1961; The Scottish Acts, 1951 and 1965; The Northern Ireland Water Act 1972.
Under these Acts it is an offence to allow poisonous, noxious or polluting matter to enter a stream, river, watercourse or inland water which discharges into a stream. In addition, water authorities may make by-laws under these Acts to control or prevent the depositing of objectionable polluting or non-polluting matter into streams etc.

Control of Pollution Act, 1974.
Under this Act it is an offence to deposit controlled waste (defined

in the Act) on land, except under licence. The Ministry of Agriculture, Fisheries and Food publication B2198, *Guidelines for the disposal of unwanted pesticides and containers on farms and holdings*, provides advice on this subject.

The Control of Pollution Act will eventually supersede the Deposit of Poisonous Waste Act and the various Rivers Acts mentioned earlier.

The Wildlife and Countryside Act, 1981.
This Act is now in force. It consists of three Parts and Part I provides legal protection for birds, some animals, and plants. It replaces the Protection of Birds Acts 1954–67, and the Conservation of Wild Creatures and Wild Plants Act 1975, with some amendments and additional matters.

In relation to weed control, it restricts the introduction into Great Britain of certain plants. In particular, it is an offence to plant or cause to grow in the wild the following plants: giant hogweed (*Heracleum mantegazzianum*), giant kelp (*Macrocystis pyrifera*), Japanese knotweed (*Polygonum cuspidatum*) and Japanese seaweed (*Sargassum muticum*).

The Farm and Garden Chemicals Regulations, 1971.
These regulations apply to labelling, in particular the requirement to display the name of the active ingredient. An EEC directive concerning labelling is implemented by PSPS, the main feature being the provision of toxic symbols on some of the more hazardous materials.

In the UK there exist, in addition to legislation covering the safe use of pesticide products, a number of items of legislation designed to put a stop to the spread of weeds and weed seeds. The Weeds Act, 1959, places the onus on landowners to prevent specified noxious weeds growing on their land from spreading their seeds to neighbouring areas. The Act also makes provision in the event of a landowner's failure to act. The weeds specified are creeping thistle (*Cirsium arvense*), spear thistle (*Cirsium vulgare*), curled dock (*Rumex crispus*), broad-leaved dock (*Rumex obtusifolius*) and common ragwort (*Senecio jacobea*).

In addition, under the EEC there is a common set of principles which regulate the quality of crop seed, and each member country must undertake to abide by these rules, which include precise guidelines concerning the purity of seed samples. For instance, there is a maximum permitted number per sample for seeds of extraneous species, including weed species.

In the UK, quality control in relation to seed samples was instituted in 1974 by a set of Seeds Regulations constituted under the powers contained in the 1964 Plant Varieties and Seeds Act as amended by the European Communities Act 1972. These are the

Cereal Seeds Regulations, the Vegetable Seeds Regulations, the Fodder Plant Seeds Regulations, the Beet Seeds Regulations, and the Oil and Fibre Plant Seeds Regulations. These Regulations lay down minimum standards of purity for each seed crop to which they apply. If official samples from seed lots reveal that the number of impurities, which include weed seeds, is greater than the level permitted in the appropriate Seeds Regulations, then marketing of that particular seed lot may be prohibited.

No matter how many Acts Parliament may pass to legislate for safety in the use of pesticide products, whether or not these materials are used safely depends, ultimately, on the individual responsible for their application.

N.B. *Forthcoming legislation*
At the time of writing, PSPS and ACAS are still in operation pending the introduction of Government legislation covering the safety and efficacy of crop protection chemicals, which will eventually replace these schemes.

Appendix 3: Selected information sources

1. M.A.F.F. PUBLICATIONS

The following publications are available from Ministry of Agriculture, Fisheries and Food (Publications), Lion House, Willowburn Trading Estate, Alnwick, Northumberland, NE66 2PF.

B2070	Horticultural Sprayers for Small Areas
B2128	Orchard Sprayers
B2272	Guildelines for Applying Crop Protection Chemicals
L767	Farm Chemical Stores
L792	Controlled Droplet Application of Agricultural Chemicals
B2049	Pasture Improvement including Use of Herbicides. Grassland Practice No. 9
B2056	Weed Control in Grassland, Herbage Legumes and grass seed crops. Grassland Practice No. 16
*B2068	Weed Control in Oil Seed Rape
*B2252	Weed Control in Cereals (Spring and Summer)
*B2253	Weed Control in Cereals (Autumn/Winter)
*B2254	Weed Control in Sugar Beet
B2256	Weed Control in Fodder Roots and Kale and Fodder Rape
*B2260	Weed Control and Haulm Destruction in Potatoes
L280	Ragwort
L452	Wild-oats
L522	Black-grass
L777	Barren brome
L778	Growth Stages in Wheat, Oats, Rye and Barley. Key No. 1.1
HVG28	Weed Control in Vegetables
B2251	Weed Control in Horticultural Brassicas
B2255	Chemical Weed Control in Strawberries
B2262	Weed Control in Peas
B2264	Chemical Weed Control in Bush and Cane Fruit
*B2361	Top Fruit Growers Guide to the Use of Chemical Sprays
*B2362	Soft Fruit Growers Guide to the Use of Chemical Sprays
STL15	Chemical Weed Control in Top Fruit Orchards
STL20	Chemical Weed Control in Flower Crops
STL69	Weed Control in Nursery Stock Production
STL191	Hops: Chemical Weed Control
B2078	Guidelines for the Use of Herbicides on Weeds in or near Water Courses and Lakes
B2198	Guidelines for the Disposal of Unwanted Pesticides and Containers on Farms and Holdings
STL188	Vegetation Control on Uncropped Land on the Farm
RB161	Poisonous Plants in Britain – and their Effects on Animals and Man (1983) HMSO (Replaces Bulletin 161).
RB281	Diagnosis of Herbicide Damage to Crops (1981). HMSO.

(L = Leaflet; STL = Short-term leaflet; B = Booklet; RB = Reference Book; * = normally updated annually).

2. SCOTTISH COLLEGES PUBLICATIONS

The three Scottish Agricultural Colleges produce, both independently and under the auspices of COSAC (Council of the Scottish Agricultural Colleges), a variety of publications relating to weeds and weed control. In the list below, the letters C, E, N and W refer to COSAC, East, North and West College publications respectively.

Individual publications are available from the following addresses:

The Publications Unit,
The East of Scotland College of Agriculture,
School of Agriculture,
West Mains Road,
EDINBURGH, EH9 3JG.
Telephone: 031-667-1041

The Librarian,
The North of Scotland College of Agriculture,
School of Agriculture,
581 King Street,
ABERDEEN, AB9 1UD.
Telephone: 0224-40291

The Librarian,
The West of Scotland Agricultural College,
Auchincruive,
AYR, KA6 5HW.
Telephone: 0292-520 331

Publications

CP56	Control of docks (1980)
CP58	Couch grass control in arable farming (1980)
CP67	The control of whins and broom (1980)
CP68	The control of rushes (1980)
CP75	Wild oat control in arable farming (1981)
CP104	Weed control in peas and beans (1983)

Technical Notes and Leaflets

CN57	Weed control in peas and beans (1983)
EN295	Pre-harvest use of roundup in cereals (1982)
EN304	Weed control in oilseed rape (1982)
EN310	Winter cereal herbicides – 1983 (1982)
EN312	Compatibility of cereal herbicides with manganese sulphate (1983)
NN12	Survey of perennial grass weeds in cereals 1981 (1981)
NN28	Wild oats in cereals 1982 (1982)
NL49	Giant hogweed (1983)
WN137	Control of common chickweed (1981)

WN148   Control of corn marigold in cereals (1982)
WN176   Herbicides in horticulture: practice and theory (1982)
WN178   Clout (alloxydim-sodium) (1982)
WN209   Surfactant additions to glyphosate
WN214   A review of weed-wiper technology.

3. BRITISH CROP PROTECTION COUNCIL (BCPC) PUBLICATIONS

Monographs

No. 24   *Spraying Systems for the 1980s* (1980)
No. 25   *Decision Making in the Practice of Crop Protection*, edited by R. B. Austin (1982)
No. 26   *Management of Vegetation*, edited by J. M. Way (1983)

Occasional Publications

No. 2   *An Explanation of the Decimal Code for the Growth Stages of Cereals*, edited by D. R. Tottman and R. J. Makepeace (1982).
No. 3   *The Influence of the Weather on the Efficiency and Safety of Pesticide Application – the Drift of Herbicides*, edited by J. G. Elliott and B. J. Wilson (1983).

Conference Proceedings

10th International Congress of Plant Protection (1983)
BCPC also publishes the *Proceedings* of the bi-annual British Crop Protection Conference – Weeds, as referenced in the text.

BCPC publications are available from:

BCPC Publications Sales,
'Shirley',
Cradley,
MALVERN,
Worcs.,
WR13 5LP.

4. SCIENTIFIC JOURNALS (MONTHLY)

(i) *Weed Research* – includes both fundamental and applied research. Frequent items of UK interest.
(ii) *Weed Science and Pesticide Science*. USA-based, important sources of background information.
(iii) *Weed Abstracts* – digest of papers, reports and other items relating to weed biology and control from throughout the world.

5. OTHER PERIODICALS

    (i) *Arable Farming* (monthly) – regular coverage of current weed and weed control developments.

   (ii) *Farmers Weekly* – weed control articles and occasional supplements.

  (iii) *Scientific Horticulture* (annually) – Journal of the Horticultural Education Association.

  (iv) *The Grower* (weekly) – mainly directed towards commercial horticulture.

  (vi) *Gardenrs Chronicle and Horticultural Trades Journal* (weekly) – amenity and general horticulture.

 (vii) Company journals – the 'house magazines' of major companies such as Shell (*Shell Agriculture*), I.C.I. (*Endeavour*), F.B.C. (*Agrospray*) and others are useful sources of information on specific developments.

Appendix 4: Glossary of Plant Names

The derivation of English and Latin names is as indicated in the Preface. In a few instances, both in the text and in the glossary, preference has been given to certain weed names which are in common use and which may differ from the above (e.g. 'onion-couch' for 'false oat-grass'); in such cases the 'official' name is also given. Only one English name has been given in each case, except where two (or more) are considered more or less equally valid (e.g. bishopweed/ground elder). In the interests of fluency and brevity, we have avoided the rigid use in the text of such prefixes as 'common' and 'field', and have adopted the terms 'wild oats', 'black grass' and ryegrass as standard.

The following abbreviations have been used:

A –   annual, mainly or wholly spring-germinating
WA –   annual, with a significant autumn/winter germination
E –   ephemeral, germinating ± throughout the year and having
           > one generation/year
B –   biennial
P –   perennial
WP –   woody perennial
Aq.P. –   aquatic perennial.

Amsinckia (Fiddleneck)	*Amsinckia intermedia*	A
Annual knawel	*Scleranthus annuus*	A
Arrowhead	*Sagittaria sagittifolia*	Aq.P
Basil thyme	*Acinos arvensis*	A
Bent, black	*Agrostis gigantea*	P
common	*A. tenuis*	P
creeping	*A. stolonifera*	P
Betony	*Betonica officinalis*	P
Bishopweed (Ground elder)	*Aegopodium podograria*	P *
Black-bindweed	*Polygonum convolvulus*	A
Black-grass	*Alopecurus myosuroides*	WA
Black medick	*Medicago lupulina*	P *
Black nightshade	*Solanum nigrum*	A
Bracken	*Pteridium aquilinum*	P
Bristle Oat	*Avena strigosa*	A
Brome, barren (Sterile brome)	*Bromus sterilis*	WA/B
meadow	*B. commutatus*	A/B
rye	*B. secalinus*	A/B
soft	*B. mollis*	A/B
Broom	*Sarothamnus scoparius*	WP
Bulrush	*Typha latifolia*	Aq.P.
Buttercup, bulbous	*Ranunculus bulbosus*	P
corn	*R. arvensis*	WA
creeping	*R. repens*	P
meadow	*R. aeris*	P

Canadian waterweed (Canadian pondweed)	*Elodea canandensis*	Aq.P.
Canary-grass	*Phalaris paradoxa*	WA
Cat's-ear	*Hypochaeris radicata*	P
Charlock	*Sinapis arvensis*	A
Chickweed, common	*Stellaria media*	E
Cleavers	*Galium aparine*	WA
Cocksfoot	*Dactylis glomerata*	P
Coltsfoot	*Tussilago farfara*	P
Comfrey	*Symphytum species*	P
Common couch (Twitch)	*Agropyron repens*	P
Common field-speedwell (Buxbaum's speedwell)	*Veronica persica*	WA
Common knapweed	*Centaurea nigra*	P
Common mallow	*Malva sylvestris*	A
Common orache	*Atriplex patula*	A
Corn chamomile	*Anthemis arvensis*	A
Corncockle	*Agrostemma githago*	A
Cornflower	*Centaurea cyanus*	A
Corn gromwell	*Lithospermum arvense*	WA
Corn marigold	*Chrysanthemum segetum*	A
Corn mint	*Mentha arvensis*	A
Corn spurrey	*Spergula arvensis*	A
Cowbane	*Cicuta virosa*	Aq.P.
Crane's-bill, dove's-foot	*Geranium molle*	A
cut-leaved	*G. dissectum*	A
Creeping soft-grass	*Holcus mollis*	P
Daisy, common	*Bellis perennis*	P *
oxeye	*Chrysanthemum leucanthemum*	P
Dandelion	*Taraxacum officinale*	P
Darnel	*Lolium temulentum*	A
Dead-nettle, red	*Lamium purpureum*	WA
Dock, broad-leaved	*Rumex obtusifolius*	P
curled	*R. crispus*	B/P
Duckweed, common	*Lemna minor*	Aq.P *
ivy-leaved	*L. trisulca*	Aq.P *
Fat-hen	*Chenopodium album*	A
Fescue, red	*Festuca rubra*	P
tall	*F. gigantea*	P
Field bindweed	*Convolvulus arvensis*	P
Field cow-wheat (Poverty-weed)	*Melampyrum arvense*	A
Field horsetail	*Equisetum arvense*	P *
Field madder	*Sherardia arvensis*	A
Field pennycress	*Thlaspi arvense*	A
Field woundwort	*Stachys arvensis*	A
hedge	*S. sylvatica*	P
Forget-me-not	*Myosotis arvensis*	WA
Fumitory	*Fumaria officinalis*	A

Gallant soldier	*Galinsoga parviflora*	A *
Gorse (whiny)	*Ulex europaeus*	WP
Greater celandine	*Chelidonium majus*	P
Groundsel, common	*Senecio vulgaris*	E *
sticky	*S. viscosus*	E
Hairy bitter-cress	*Cardamine hirsuta*	A *
Hawthorn	*Crataegus monogyna*	WP
Hedge-parsley, spreading	*Torilis arvensis*	A
upright	*T. japonica*	A
Hemlock	*Conium maculatum*	P
Hemlock water dropwort	*Oenanthe crocata*	Aq.P.
Hemp-nettle, common (Day-nettle)	*Galeopsis tetrahit*	A
large flowered	*G. speciosa*	A
hoary cress (Thanet cress; Hoary pepperwort)	*Cardaria draba*	P
Hogweed, common	*Heracleum sphondylium*	P
giant	*H. mantegazzianum*	P
Japanese knotweed (Japanese bamboo)	*Polygonum cuspidatum*	P *
Knotgrass	*P. aviculare*	A
Larkspur	*Delphinium ambiguum*	A
Lesser celandine	*Ranunculus ficaria*	P
Lesser snapdragon	*Misopates orontium*	A
Loose silky-bent	*Apera spica-venti*	A
Many-seeded goosefoot (Allseed)	*Chenopodium polyspermum*	A *
Marsh foxtail	*Alopecurus geniculatus*	P
Mayweed, scented	*Matricaria recutita*	A
scentless	*Tripleurospermum maritimum subsp. inodorum*	A
stinking	*Anthemis cotula*	A
Meadow-grass, annual	*Poa annua*	E
rough-stalked	*P. trivialis*	P
smooth-stalked	*P. pratensis*	P
Mouse ear (chickweed)	*Cerastium holosteiodes*	A
Mouse-tail	*Myosurus minimus*	A
Mustard, black	*Brassica nigra*	A
garlic	*Alliaria petiolata*	B
treacle	*Erysimum cheiranthoides*	A/B *
Nettle, common (stinging nettle)	*Urtica dioica*	P
annual (small nettle)	*U. urens*	A *
Night-flowering catchfly	*Silene noctiflora*	A
Nipplewort	*Lapsana communis*	A

Onion-couch (False oat-grass)	*Arrhenatherum elatius*	P
Pale persicaria	*Polygonum lapathifolium*	A
Pansy, field	*Viola arvensis*	A
wild	*V. tricolor*	WA
Parsley, cow (Keck)	*Anthriscus sylvestris*	P
Parsley-piert	*Aphanes arvensis*	WA
Pheasant's eye	*Adonis annua*	A
Pineappleweed	*Matricaria matricarioides*	A
Plantain, greater	*Plantago major*	P
hoary	*P. media*	P
ribwort	*P. lanceolata*	P
Poppy, common (corn)	*Papaver rhoeas*	WA
long-headed	*P. dubium*	A
prickly (long-headed)	*P. argemone*	A
rough	*P. hybridum*	A
Procumbent pearlwort	*Sagina procumbens*	P *
Procumbent yellow sorrel	*Oxalis corniculata*	AP *
Ragwort, common	*Senecio jacobaea*	B/P
marsh	*S. aquaticus*	B
Oxford	*S. squalidus*	A
Red bartsia	*Odontites verna*	P
Rhododendron	*Rhododendron ponticum*	WP *
Runch (Wild radish)	*Raphanus raphanistrum*	A
Rush, compact	*J. conglomeratus*	P
hard	*J. inflexus*	P
heath	*J. squarrosus*	P
jointed	*J. articulatus*	P
sharp-flowered	*J. acutiflorus*	P
soft (Common rush)	*J. effusus*	P
Ryegrass, Italian	*Lolium multiflorum*	B
perennial	*L. perenne*	P
Scarlet pimpernel	*Anagallis arvensis*	A/P
Shepherd's needle	*Scandix pecten-veneris*	A
Shepherd's purse	*Capsella bursa-pastoris*	WA
Silverweed	*Potentilla anserina*	P
Small-flowered balsam	*Impatiens parviflora*	A
Sorrel, common	*Rumex acetosa*	P
sheep's	*R. acetosella*	P *
Sow-thistle, perennial	*Sonchus arvensis*	P
prickly	*S. asper*	A
smooth (milk)	*S. oleraceus*	A
Speedwell, grey	*Veronica polita*	
ivy-leaved	*V. hederifolia*	WA
slender	*V. filiformis*	P *
wall	*V. arvensis*	P
Spurge, petty	*Eurphorbia peplus*	A *
sun	*E. helioscopia*	A *
Starthistles	*Centaurea species*	B/P
Lesser swine-cress	*Coronopus didymus*	A

Tare, hairy	*Vicia hirsuta*	A
smooth	*V. tetrasperma*	A
Thistle, creeping	*Cirsium arvense*	P
marsh	*C. palustre*	B
musk	*Carduus nutans*	B
spear	*Cirsium vulgare*	B
welted	*Carduus acanthoides*	B
Thorow-wax	*Bupleurum rotundifolium*	A
Timothy	*Phleum pratense*	P
Toadflax, common	*Linaria vulgaris*	P
Tufted hair-grass	*Deschampsia cespitosa*	P
Wild barley (Wall barley)	*Hordeum murinum*	A
White clover	*Trifolium repens*	P *
Wild-oat, spring	*Avena fatua*	A
winter	*A. ludoviciana*	WA
Wild onion	*Allium vineale*	P
Water crowfoot	*Ranunculus aquatilis*	Aq.P.
Water plantain	*Alisma plantago-aquatica*	Aq.P.
Willowherb, hairy	*Epilobium hirsutum*	P
rosebay	*Chamaenerion angustifolium*	P
Yellow water-lily	*Nuphar lutea*	Aq.P *
Yew	*Taxus baccata*	WP
Yorkshire fog	*Holcus lanatus*	P

* indicates species particularly associated with horticultural situations.

Recent name changes

A number of the scientific names in the above list have been revised in recent years, although many of the new names have yet to come into common use. The source for these changes is *Flora Europaea*, ed. T. G. Tutin *et al.*, in 5 vols., 1964–80, at The University Press, Cambridge.

The relevant alternatives are given below, with the old names shown first:

Agropyron repens	*Elymus repens (L.) Gould*
Agrostis tenuis	*Agrostis capillaris L.*
Avena ludoviciana	*A. sterilis ssp.*
	ludoviciana (Dur.) Nyman
Bromus mollis	*B. hordeaceus L.*
Cerastium holosteoides	*C. fontanum Baumg*
Chamaenerion angustifolium	*Epilobium angustifolium*
Chrysanthenum leucanthemum	*Leucanthemum vulgare Lam*
Gnaphalium uliginosum	*Filaginella uliginosa (L.) Opis*
Lithospermum arvense	*Buglossoides arvensis (L.) John*
Matricaria matricarioides	*Chamomilla suaveolens (Pursh) Rydb*
Matricaria recutita	*Chamomilla recutita (L.) Rausch*
Polygonum convolvulus	*Bilderdykia conovolvulus (L.) Dumort*
Polygonum cuspidatum	*Reynoutria japonica Houtt*
Tripleurospermum maritimum	*Matricaria perforata Merat*
ssp inodorum	

Selected Bibliography

ANON (1970), 'The basic requirements for applying agrochemicals', *Fisons Agricultural Technical Information*, Autumn, 1970, 7–11.

ANON (1981), 'Give trees a chance', *G.C. and H.T.J.*, April 24, 1981, 24–8.

ANON (1981), 'Tested for safety', *G.C. and H.T.J.*, October 23, 1981, 15–17.

ANON (1981), 'Seeds and weeds', *G.C. and H.T.J.*, November 20, 1981, 73–4.

ANON (1982), 'An idea that gels', *G.C. and H.T.J.*, January 1, 1982, 25.

ANON (1982), 'A step forward', *G.C. and H.T.J.*, January 22, 1982, 25.

ANON (1982), 'British climate reduces plastic advantages', *Grower* February 25, 1982, 15–16.

ANON (1982), 'In park and garden', *G.C. and H.T.J.*, March 12, 1982, 13–14.

ANON (1982), 'The amenity armoury', *G.C. and H.T.J.*, March 12, 1982, 35–7.

ATKIN, J. C. and TURNER, M. T. F. (1982), 'Control of blackgrass and wild oats with a single granular application', *Proc. 1982 Br. Crop Prot Conf. – Weeds*, 637–44.

ATKINSON, D. and HERBERT, R. F. (1978), 'Long-term comparison of the effects of soil management', *Rep. E. Malling Res. Stn. for 1977, 53–4.*

ATKINSON, D. and HERBERT, R. F. (1979), 'Effects on the soil with particular reference to orchard crops', *Ann. appl. Biol., 91*, 125–9.

ATKINSON, D. and WHITE, G. C. (1976), 'Soil management with herbicides – the response of soils and plants', *Proc. 1976 Br. Crop. Prot. Conf. – Weeds*, 873–84.

AUDUS, L. J. (ed.) (1976), *Herbicides: physiology, biochemistry, ecology*, 2nd edition, Academic Press, London, New York and San Francisco (2 vols).

AYRES, P. (1980), 'The implications of high-speed, low-volume spraying . . . in winter cereals', *Proc. Br. 1982 Crop Prot. Conf. – Weeds*, 687–93.

AYRES, P. (1982), 'The effect of sequential reduced rates of diclofop-methyl and isoproturon on the control of *Alopecurus myosuroides* in winter wheat', *Proc. 1982 Br. Crop Prot. Conf. – Weeds*, 645–52.

BAKER, H. G. (1974), 'The evolution of weeds', *Ann. Rev. Ecol. Syst. 5*, 1–24.

BAILEY, R. J., PHILLIPS, M., HARRIS, P. and BRADFORD, A. (1982), 'The results of an investigation to determine the optimum drop size and volume of application for weed control with spinning disc applicators' *Proc. Br. 1982 Crop Prot. Conf. – Weeds*, 995–1000.

BALDWIN, J. H. (1979), 'The chemical control of wild oats and blackgrass', *A.D.A.S. Quarterly Review*, *33*, 69–101.

BALDWIN, J. H. (1981), 'A review of trials by the Agricultural Development and Advisory Service on *Alopecurus myosuroides*', *Proc. Grass Weeds in Cereals in the United Kingdom Conf.*, 1981, 197–205.

BIGGIN, P. (1980), 'Weed control in conifer transplant lines', *Proceedings Weed Control in Forestry Conference*.

BRADFORD, A. M. and SMITH, J. (1982), 'Annual broad-leaved weed control in winter cereals – ADAS 1982 results', *Proc. 1982 Br. Crop Prot. Conf. – Weeds*, 539–44'.

BRAY, W. (1983), 'Sugar beet weed control', *Br. Sugar Beet Review*, *51*, Spring 1983, 3–5.

BUDD, E. G. (1981), 'Survey, dormancy and life-cycle of *Bromus sterilis* (Barren Brome) in cereals, with particular reference to spring barley', *Proc. Grass Weeds in Cereals in the United Kingdom Conf.* 1981, 23–30.

CARTER, A. R. (1978), 'The use of herbicides in hardy ornamental crops', *Proc. 1978 Br. Crop Prot. Conf. – Weeds*, 883–8.

CHAMBERS, D. (1981), 'A look at electrostatic spraying'. *Agrospray*, No. 4, 1981, FBC Ltd, 16–17.

CHANCELLOR, R. J. (1979), 'The long-term effects of herbicides on weed populations', *Ann. appl. Biol.*, *91*, 141–4.

CHANCELLOR, R. J. and DAVISON, J. G. (1973), 'How to manage garden weeds', in Green, P. S. (ed.), *Plants: wild and cultivated*, Bot. Soc. Br. Isles, 27–37.

CHANCELLOR, R. J. and PETERS, N. C. B. (1976), 'Competition between wild oats and crops', in PRICE-JONES, D. (ed.) *Wild oats in World Agriculture*, Agric. Res. Council, London, 99–113.

CLAY, D. V. (1981), 'Weed control – where next?', *Grower*, April 23, 1981, 64–70.

CLAY, D. V. (1981), Evaluating residual orchard herbicides, *Grower*, October 15, 1981, 36–43.

COOPER, F. B. (1982), 'Experiences in controlling *Bromus mollis* in permanent swards', *Proc. 1982 Br. Crop Prot. Conf. – Weeds*, 381–5.

COURTNEY, A. D. and JOHNSTON, R. T. (1982), 'The influence of competitive stress on the components of yield in spring barley', *Aspects of Applied Biology*, *1*, 1982; Broad-leaved Weeds and their Control in Cereals, 239–46.

CUSSANS, G. W. (1970), 'A study of the competition between *Agropyron repens* and spring barley, wheat and field beans', *Proc. 10th (1970) Br. Crop Prot. Conf. – Weeds*, 337–44.

CUSSANS, G. W. (1978), 'The problem of volunteer crops and some possible means of their control', *Proc. 1978 Br. Crop Prot. Conf. – Weeds*, 915–21.

CUSSANS, G. (1980), 'Weeds in cereals and their control', *Span*, 23, 1, 30–2.

CUSSANS, G. W. (1981), 'Weed Control in Cereals – a Long-term View', *Proc. Grass Weeds in Cereals in the United Kingdom Conf.*, 1981, 355–61.

CUSSANS, G. W., MOSS, S. R.., POLLARD, F. and WILSON, B. J. (1979), 'Studies of the effects of tillage on annual weed populations', *Proc. EWRS Symp. The Influence of Different Factors on the Development and Control of Weeds*, 1979, 115–22.

DAVIES, A. (1982), 'All under Control', *G.C. and H.T.J.*, March 19, 1982, 29–37.

DAVISON, J. G. (1976), 'The effect of weeds on field-grown nursery stock', *ARC Research Review 2*, (3), 76–9.

DAVISON, J. G. (1978), 'Weed control in fruit crops – what's needed', *Proc. 1978 Br. Crop Prot. Conf. – Weeds*, 897–904.

DAVISON, J. G. (1981), 'Black plastics benefit young trees', *Grower*, October 15, 1981, 44–6.

DAVISON, J. G. and BAILEY, J. A. (1976), 'The response of four varieties of strawberry to 2,4-D applied on five dates in the year of planting, *Proc. 13th Br. Crop Prot. Conf. – Weeds* (1976), 273–80.

DAVISON, J. G. and BAILEY, J. A. (1979), 'Black polythene for weed control in young fruit and other perennial crops', *ARC Research Review*, British Growers Look Ahead Issue, 11–14.

DAVISON, J. G. and BAILEY, J. A. (1980), 'The effect of weeds on the growth of a range of nursery stock species planted as lines and grown for two seasons', *Proceedings Weed Control in Forestry Conference, 1980*, 13–20.

DAVISON, J. G. and CLAY, D. V. (1972), 'The persistence of soil-applied herbicides', *Span*, *15*, (2), 68–71.

DAVISON, J. G. and CLAY, D. V. (1978), 'Controlling weeds "with care" ', *Grower*, January 12, 1978, 75–6.

DAVISON, J. G. and ROBERTS, H. A. (1976), 'Influence of changing husbandry on weeds and weed control in horticulture, *Proc. 1976 Br. Crop Prot. Conf. – Weeds*, 1009–1017.

EAGLE, D. J. (1982), 'Hazard to adjoining crops from vapour drift of phenoxy herbicides applied to cereals', *Proc. 1982 Br. Crop Prot. Conf. – Weeds*, 33–41.

EDWARDS, C. A. and STAFFORD, C. J. (1979), 'Interactions between herbicides and the soil fauna', *Ann. appl. Biol. 91*, 132–7.

ELLIOTT, J. G. (1978), The economic objective of weed control in cereals, *Proc. 1978 Br. Crop. Prot. Conf. – Weeds*, 829–40.

ELLIOTT, J. G. (1980), 'The economic significance of weeds in the harvesting of grain', *Proc. 1980 Br. Crop. Prot. Conf. – Weeds*, 787–97.

ELLIOTT, J. G. (1981), 'Back to a four-course rotation to beat problem weeds', *Arable Farming*, Jan. 1981, 64.

ELLIOTT, J. G., CHURCH, B. M., HARVEY, J. J., HOLROYD, J., HULLS, R. H. and WATERSTON, H. A. (1979), 'Survey of the presence and methods of control of wild-oat, blackgrass and couch grass in cereal crops in the United Kingdom during 1977', *J. Ag. Sci. Camb.*, 92, 617–34.

ERNLE, Lord (1961), *English Farming Past and Present*. 6th Ed., Heinemann, London.

EVANS, J. G. (1976). *The Environment of Early Man in the British Isles*, Book Club Associates/Paul Elek.

EVANS, T. A. and STANBURY, M. A. (1982), 'Selective control of *Ulex gallii* and *Ulex europaeus* (gorse) in a *Calluna vulgaris* (heather) dominant hill sward', *Proc. 1982 Br. Crop Prot. Conf. – Weeds*, 415–18.

FENTON, A. (1976), *Scottish Country Life*, John Donald Publishers, Edinburgh.

FINNEY, J. R. (1979), 'Future weed problems and their control', *Ann. appl. Biol. 91*, 144–6.

FLETCHER, W. W. and KIRKWOOD, R. C. (1982), *Herbicides and Plant Growth Regulators*, Granada Publishing Ltd.

FORSYTH, A. A. (1979), *British Poisonous Plants*, M.A.F.F. Bulletin 161, HMSO, London.

FREE, J. B. (1968), 'Dandelion as a competitor to fruit trees for bee visits', *Journal of Applied Ecology*, *5*, (1), 169–78.

FROUD-WILLIAMS, R. J. (1981), 'Germination behaviour of *Bromus* spp. and *Alopecurus myosuroides*', *Proc. Grass Weeds in Cereals in the United Kingdom Conf.*, 1981, 31–40.

FROUD-WILLIAMS, R. J. and CHANCELLOR, R. J. (1982), 'A survey of grass weeds in cereals in central southern England', *Weed Research*, *22*, 163–71.

FRYER, J. D. and MAKEPEACE, R. J. (Eds.) (1978), *Weed Control Handbook Volume II: Recommendations*, 8th Edn., Blackwell Scientific Publications.

GARDNER, A. J. (1983), *National Agricultural Conf. – Better Spraying*, Jan. 1983.

GLADDERS, P. and MUSA, T. M. (1982), 'Effects of several herbicides on diseases of oilseed rape', *Proc. 1982 Br. Crop Prot. Conf. – Weeds*, 115–22.

GLOB, P. V. (1969), *The Bog People*, Faber and Faber.

GODWIN, H. (1956), *The History of the British Flora*, Cambridge University Press, Cambridge.

GOLDSWORTHY, J. A. and DRUMMOND, D. J. (1981), 'Control of soft brome (*Bromus mollis*) in established grassland of Scotland with ethofumesate', *Proc. Crop. Prot. in Northern Britain Conf.*, 1981, 225–30.

GREAVES, M. P. (1979), 'Long-term effects of herbicides on soil micro-organisms', *Ann. appl. Biol.*, *91*, 129–32.

GROSSBARD, E. (1971), 'Do herbicides affect micro-organisms of the soil?', *Weed Research Organisation Fourth Report 1969–1971*, 72–83.

HABESHAW, D. (1982), 'Cutting down on herbicide damage', *The Scottish Farmer*, April 17, 1982, 54–5.

HADDOW, B. C., ICKERINGILL, D., and MONCORGE, J. M. (1978), 'L-Flamprop-isopropyl. A versatile wild oat herbicide with economic benefits for use in wheat and barley', *Proc. 1978 Br. Crop Prot. Conf. – Weeds*, 23–30.

HAFLINGER, E. and BRUN-HOOL, J. (1971), *Weed Communities of Europe*, Ciba-Geigy, Basle.

HAGGER, R. J. and KIRKHAM, F. W. (1981), 'Selective herbicides for establishing weed-free grass', I. *Weed Research*, *21*, 141–51.

HANCE, R. J. (1979), 'Herbicide persistence and breakdown in soil in the long term', *Ann. appl. Biol.*, *91*, 137–41.

HANCE, R. J. (ed.) (1980), *Interactions Between Herbicides and the Soil*, Academic Press, London.

HARKESS, R. D. and FRAME, J. (1981), 'Weed grasses in sown grassland, with special reference to soft brome (*Bromus mollis* L.)', *Proc. Crop Prot. in Northern Britain Conf.*, 1981, 231–6.

HARVEY, N. (1980), *The Industrial Archaeology of Farming in England and Wales*, Batsford.

HEWSON, R. T. and ROBERTS H. A. (1971), 'The effect of weed removal at different times on the yield of bulb onions', *J. Hort. Sci. 46*, 471–83.

I.C.I. (1981), *Growing Cereals*, 39–44.

JOHNSTONE, D. R. (1978), 'The influence of physical and meteorological factors on the deposition and drift of spray droplets of controlled size', *Proc. Br. Crop Prot. Coun. Monograph, No. 22*, Controlled Droplet Application, 43–57.

JONES, A. (1981), 'Safe and sound', *G.C. and H.T.J.*, May 8, 1981, 25–6.

KEARNEY, P. C. and KAUFMAN, D. D. (eds.) (1976), *Herbicides: Chemistry, Degradation, Mode of Action*, Dekker, New York.

KEARNEY, P. C., NASH, R. G. and ISENSEE, A. R. (1969), 'Persistence of pesticide residues in soils', in Miller, M. W. and Berg, G. G. (eds), *Chemical Fallout*, Springfield, Illinois.

KING, J. M. (1972), 'Objectives of weed control in vegetables. The effects of weeds and herbicides on the growth and yield of vegetable crops', *Proc. 11th Br. Weed Cont. Conf.*, 925–31.

KING, J. M. (1980), 'Methods of testing the reaction to herbicides of varieties of peas, broad beans and dwarf beans and the practical value of the results', *Proc. 1980 Br. Crop Prot. Conf. – Weeds*, 453–60.

KIRKHAM, F. W. and HAGGER, R. J. (1982), 'Selective herbicides for establishing weed-free grass', II. *Weed Res.*, *22*, 57–68.

LAKE, J. R., GREEN, R., TOFTS, M. and DIX, A. J. (1982), 'The effect of an aerofoil on the penetration of charged spray into barley', *Proc. 1982 Br. Crop Prot. Conf. – Weeds*, 1009–1016.

LARGE, J. W. (1981), 'Wild oat control on a farm scale', *Proc. Grass Weeds in Cereals in the United Kingdom Conf.*, 1981, 363–6.

LAWRENCE, D. C. (1980), 'Aids to swathe matching in tractor spraying', *Br. Crop Prot., Coun. Monogr. No. 24 – Spraying Systems for the 1980s*, 215–27.

LAWSON, H. M. (1974), 'The effect of weeds on fruit and ornamental crops', *Proc. 12th Br. Weed Cont. Conf.*, 1159–1169.

LAWSON, H. M. (1979), 'The influence of herbicides and crop management on weed control problems in raspberries', *Proc. Symposium on the influence of different factors on the development and control of weeds*, European Weed Research Society, 201–8.

LAWSON, H. M. (1980), 'Biology and control of raspberry suckers', *Scientific Horticulture*, *31*, (4), 101–5.

LAWSON, H. M. (1983), 'Competition between annual weeds and vining peas grown at a range of densities: effects on the crop', *Weed Research*, *23*, 27–38.

LAWSON, H. M. and RUBENS, T. G. (1970), 'Experiments on the control of perennial weeds in established raspberry plantations.' *Proc. 11th Br. Weed Cont. Conf.*, 760–767.

LAWSON, H. M. and WISEMAN, J. S. (1976), 'Weed control in spring planted raspberries', *Weed Research*, *16*, 155–62.

LAWSON, H. M. and WISEMAN, J. S. (1979), 'Effects of raspberry suckers, growing in the alleys between rows, on cane and fruit production in a non-cultivated plantation', *Hort. Res.*, *19*, 63–74.

LAWSON, H. M. and WISEMAN, J. S. (1982), 'Relative tolerance of calabrese and swede to three specific graminicides', *Proc. 1982 Br. Crop Prot. Conf. – Weeds*, 927–30.

LOCKETT, P. M. and ROBERTS, H. A. (1976), 'Weed seed populations', *Report National Vegetable Research Station for 1975*, 114.

LUTMAN, P. J. W. (1980), 'A review of techniques that utilize height differences between crops and weeds to achieve selectivity', *Br. Crop Prot. Coun. Monogr. No. 24 – Spraying Systems for the 1980s*, 291–7.

MAKEPEACE, R. J. (1982, a), 'A review of broad-leaved weed problems in spring cereals', *Aspects of Applied Biology*, *1*, 1982: Broad-leaved weeds and their Control in Cereals, 103–8.

MARSHALL, E. J. P. (1981), 'Managing rural amenity sites with chemicals', *Weed Research Organisation, Ninth Report, 1980–81*, 71–7.

MARSHALL, G. (1979), 'Controlled drop application (CDA) of herbicides', *West of Scotland Agricultural College Technical Note Number 45*, January 1979.

MARSHALL, G. (1981), 'Recent developments in herbicide application techniques', *West of Scotland Agricultural College Technical Note Number 133*, March 1981.

MARSHALL, P. F. (1981), 'Nozzle-nous', *Power Farming*, April 1981, 14–15.

MARTINDALE, J. F. and LIVINGSTONE, D. B. (1982), 'Chemical control of *Phalaris paradoxa* in winter cereals', *Proc. 1982 Br. Crop Prot. Conf.-Weeds*, 671–8.

MATTHEWS, G. A. (1979), 'Controlled droplet application', in *Pesticide Application Methods*, Longman, London.

MERRITT, C. R. (1980), 'The influence of application variables on the biological performance of foliage-applied herbicides', *Proc. Br. Crop Prot. Coun. Monogr. No. 24 – Spraying Systems for the 1980s*, 35–43.

MIDGLEY, S. J. (1982), 'Effects of surfactants on phenoxyalkanoic herbicides: a preliminary report', *Aspects of Applied Biology 1, 1982*: Broad-leaved Weeds and their Control in Cereals, 193–200.

MOSS, S. R. (1979), 'Black-grass – a threat to winter cereals', *Technical leaflet*, ARC Weed Res. Organisation, *15*, pp. 3.

MOSS, S. R. (1981), 'The response of *Alopecurus myosuroides* during a four year period to different cultivation and straw-disposal systems', *Proc. Grass Weeds in Cereals in the United Kingdom Conf.*, 1981, 15–21.

MOSS, S. R. (1983), 'The production and shedding of *Alopecurus myosuroides* Huds, seeds in winter cereal crops', *Weed Research*, *23*, 45–51.

MUNDY, E. J. (1975), 'Herbicides have an edge over cultivations', *Arable Farming*, June 1975, 41–2.

MURANT, A. F. and LISTER, R. M. (1976), 'Seed transmission in the ecology of nematode-borne viruses', *Ann. appl. Biol. 59*, 63–76.

MURRAY, R. B. (1982), 'Herbicides in Horticulture: Practice and theory', *West of Scotland Agricultural College Technical Note Number 176*, October, 1982.

MURRAY, R. B. and WOODS, A. M. (1981), 'A survey of current herbicide usage by Local Authorities in Scotland', *HEA Weed Control Meeting*, The West of Scotland Agricultural College.

O'LEARY, T. E. (1973), 'Broad-leaved weed infestations in cereals', *Fisons Agricultural Technical Information* (Agtec.) Spring 1973, 37–41.

ORSON, J. H. (1982, a), 'Annual broad-leaved weed control in winter wheat and winter barley, autumn and spring treatments compared', *Aspects of Applied Biology 1*, 1982, 43–51.

ORSON, J. H. (1982, b), 'The control of *Agropyron repens* pre-harvest of wheat and barley with the isopropylamine salt of glyphosate – ADAS Results 1980 and 1981', *Proc. 1982 Br. Crop Prot. Conf. – Weeds*, 653–60.

OSWALD, A. K. (1982), 'Two surveys of the potential use of a technique for controlling tall weeds in grassland', *Proc. 1982 Br. Crop. Prot. Conf. – Weeds*, 387–94.

PARFITT, R. I., STINCHCOMBE, G. R. and STOTT, K. G. (1980), 'The establishment and growth of windbreak trees in polythene mulch, straw mulch and herbicide maintained bare soil', *Proc. 1980 Br. Crop Prot. Conf. – Weeds*, 739–46.

PETERS, N. C. B. and WILSON, B. J. (1983), 'Some studies on the competition between *Avena fatua* L. and spring barley. II. Variation of *A. fatua* emergence and development and its influence on crop yield', *Weed Research*, *23*, 305–11.

PHILLIPSON, A. (1974), 'Survey of the presence of wild oats and black-grass in parts of the United Kingdom', *Weed Res.*, *14*, 123–35.

POTTS, M. J. (1973), 'The biology of wild oats', *Proc. Wild Oats Symposium*, The West of Scotland Agricultural College, 5–21.

RACKHAM, O. (1976), *Trees and Woodland in the British Landscape*, J. M. Dent & Sons.

ROBERTS, H. A. (1970), 'Viable weed seeds in cultured soils', *Rept. Natn. Veg.*

Res. Stn. for 1969, 25–38.

ROBERTS, H. A. (1981), 'Seed banks in soils', *Ann. appl. Biol.*, 6, 1–55.

ROBERTS, H. A. (ed.) (1982), *The Weed Control Handbook*, 7th ed., Vol. 1, Blackwell Scientific Publications, Oxford.

ROBERTS, H. A. (1983), 'Weed seeds in horticultural soils', *Scientific Horticulture*, 34, 1–11.

ROBERTS, H. A. and POTTER, M. E. (1980), 'Emergence patterns of weed seedlings in relation to cultivation and rainfall', *Weed Research*, 20, 377–86.

ROBINSON, D. W. (1964), 'Weed control in horticulture – a review', *Proc. 7th Br. Weed Cont. Conf.*, 994–1009.

ROBINSON, D. W. (1978), 'The challenge of the next generation of weed problems', *Proc. 1978 Br. Crop Prot. Conf. – Weeds*, 799–823.

ROWE-DUTTON, P. (1976), 'Mulching is important', *The Garden*, 101, 135–9.

RULE, J. S. (1981), 'Grass weed problems old and new', *Arable Farming*, Oct. 1981, 63–4.

RUTHERFORD, I. (1980), 'Vehicle design and performance for pesticide application', *Proc. Br. Crop Prot. Coun. Monogr. No. 24, Spraying Systems for the 1980s*, 185–98.

SCRAGG, E. B. (1980), 'Cost-effective weed control in spring barley in the north of Scotland', *Proc. 1980 Br. Crop Prot. Conf. – Weeds*, 69–75.

SCRAGG, E. B. and KILGOUR, D. W. (1981), 'A survey of wild oats in barley and wheat in N. E. Scotland 1975–80', *Proc. Crop Prot. in Northern Britain Conf.*, 1981, 57–62.

SELMAN, M. (1970), 'Population dynamics of *Avena fatua* in continuous spring barley', *Proc. 1970 Br. Crop Prot. Conf. – Weeds*, 1176–1188.

SHEPPARD, B. W., PASCAL, J. A., RICHARDS, M. C. and GRANT, H. (1982), 'The control of *Agropyron repens* by the pre-harvest application of glyphosate', *Proc. 1982 Br Crop Prot. Conf. – Weeds*, 953–61.

SMITH, J. (1983), 'Sugar beet weed control', *Br. Sugar Beet Review*, 51, Spring 1983, 24–5.

SMITH, J. and FINCH, R. J. (1978), 'Chemical control of *Avena fatua* in spring barley', *Proc. 1978 Br. Crop Prot. Conf. – Weeds*, 841–9.

STEPHENSON, G. R. and EZRA, G. (1982), 'The mode of action of herbicide safeners', *Proc. 1982 Br. Crop Prot. Conf.–Weeds*, 451–60.

SWIFT, G. (1978), 'A survey of grassland in East Scotland: Interim Report 1976–77', *East of Scotland Agric. College Technical Note 182C/A*.

TAYLOR, J. A. H. (1980), 'Herbicide usage on potatoes in Great Britain in 1980', *Proc. 1980 Br. Crop Prot.Conf.–Weeds*, 891–8.

TAYLOR, W. A. (1981), 'Controlled droplet application of herbicides', *Outlook on Agriculture*, 10, 333–6.

THOMPSON, N. (1982), 'Meteorology as an aid to Crop Protection', *Br. Crop Prot. Council Monogr. No. 25, Decision Making in the Practice of Crop Protection*, 55–63.

TOTTMAN, D. R. (1976), 'Spray timing and the identification of cereal growth stages', *Proc. 1976 Br. Crop Prot. Conf. – Weeds*, 791–800.

TOTTMAN, D. R. and MAKEPEACE, R. J. (1982), 'An Explanation of the Decimal Code for the Growth Stages of Cereals', *Occ. Pub. No. 2*. Br. Crop Prot. Council Publications, Malvern.

UPSTONE, M. E., SWANN, N. B. and STICHBURY, R. (1982), 'Control of *Alopecurus myosuroides* in U.K. by autumn application of chlorsulfuron plus methabenzthiazuron in winter wheat', *Proc. 1982 Br. Crop Prot. Conf. - Weeds*, 629–36.

WALKER, A. (1983), 'The fate and significance of herbicide residues in soil', *Scientific Horticulture, 34*, 35–47.

WALKER, A. and ROBERTS, H. A. (1983), 'Avoiding the problem of herbicide residues', *Horticulture Now*, No. 7, 38–40.

WARD, J. T. (1982), 'Weed control in winter oilseed rape', *Proc. 1982 Br. Crop Conf. - Weeds*, 97–102.

WATSON, E. (1981), 'Spray Drift' Parts I, II and III, *Grower* March 5, 1981, 33–40; *Grower* March 12, 1981, 22–6; *Grower* March 19, 1981, 24–30.

WAY, J. M. (1972), 'The use of herbicides on non-crop areas on a farm', *Br. Crop Prot. Coun. Monogr. No. 6*, 21–8.

WAY, J. M. (1974), 'The environmental consequences of using herbicides', *Proc. 12th Br. Weed Cont. Conf.*, 925–8.

WELLS, G. J. (1979), 'Annual weed competition in wheat crops: the effect of weed density and applied nitrogen', *Weed Research, 19*, 185–92.

WHITE, G. C. and HOLLOWAY, R. I. C. (1967), 'The influence of simazine or a straw mulch in the establishment of apple trees in grassed down or cultivated soil', *J. Hort. Sci.,, 42*, 377–89.

WILLIAMS, G. H. W. (1980), 'Follow-up treatments for control of *Pteridium aquilinum*', *Proc. 1980 Br. Crop Prot. Conf. - Weeds*, 423–8.

WILSON, B. J. (1981, b), 'The influence of reduced cultivations and direct drilling on the long-term decline of a population of *Avena fatua* L. in spring barley', *Weed Research, 21*, 23–8.

WILSON, B. J. and CUSSANS, G. W. (1975), 'A study of the population dynamics of *Avena fatua* L. as influenced by straw-burning, seed-shedding and cultivations', *Weed Research, 15*, 249–58.

WOODS, A. M. (1980), 'A study of herbicide usage in amenity horticulture with particular reference to current practices in Local Authorities in Scotland', B.Sc. Thesis, the University of Strathclyde, Glasgow.

ZADOKS, J. C., CHANG, T. T. and KONZAC, C. F., 'A decimal code for the growth stages of cereals', *Weed Research, 14*, 415–21.

Index